THE LEGO® MINDSTORMS® NXT® IDEA BOOK

THE LEGO® MINDSTORMS® NXT® IDEA BOOK

design, invent, and build

martijn **boogaarts**, jonathan a. **daudelin**, brian l. **davis**,

jim **kelly**, david **levy**, lou **morris**, fay **rhodes**, rick **rhodes**,

matthias paul **scholz**, christopher r. **smith**, and rob **torok**

no starch press

Printed in the United States of America

11 10 09   4 5 6 7 8 9

ISBN-10: 1-59327-150-6
ISBN-13: 978-1-59327-150-3

Publisher: William Pollock
Production Editor: Megan Dunchak and Elizabeth Campbell
Cover and Interior Design: Octopod Studios
Developmental Editor: William Pollock
Copyeditors: Nancy Sixsmith and Megan Dunchak
Compositor: Riley Hoffman
Proofreader: Susie Seefelt Lesieutre
Indexer: Nancy Guenther

For information on book distributors or translations, please contact No Starch Press, Inc. directly:

No Starch Press, Inc.
555 De Haro Street, Suite 250, San Francisco, CA 94107
phone: 415.863.9900; fax: 415.863.9950; info@nostarch.com; www.nostarch.com

Library of Congress Cataloging-in-Publication Data

The LEGO Mindstorms NXT idea book : design, invent, and build / by Martijn Boogaarts ... [et al.].
        p. cm.
    ISBN-13: 978-1-59327-150-3
    ISBN-10: 1-59327-150-6
    1. Robots--Design and construction. 2. Robots--Programming. 3. LEGO toys. I. Boogaarts, Martijn.
TJ211.L4525 2007
629.8'92--dc22
                                    2007021511

about the contributors

Martijn Boogaarts

Martijn Boogaarts (from Duiven, The Netherlands) is a freelance integration technology trainer. In 1986, he started a LEGO® "robotica" club at his school and he has since built many robots. Martijn was one of the initial organizers of LEGO WORLD, and he built several large demonstration models, including the Road-Plate-Laying Machine, a working car factory (27 RCXs), and a pinball machine. In April 2005, he contributed to the AFOL MINDSTORMS® tournament in Billund, Denmark, and later that year he was asked to join the MINDSTORMS Users Panel (MUP2). Martijn contributes to The NXT STEP blog and shares knowledge about the NXT® to show that you can build it, too. (That's his robot on the cover, from Chapter 12, "CraneBot: A Grabber Robot.")

Jonathan A. Daudelin

Jonathan A. Daudelin is a 15-year-old who's been building LEGO MINDSTORMS robots as a hobby for six years. He enjoys using CAD software to render his robots on the computer. He also helped start and was a member of a FIRST LEGO League team, Built On The Rock. In their second year of competing, he and his team won first place in the Robot Performance and Innovative Robot categories at the World Festival in the 2007 Nano Quest challenge. His team's robot achieved perfect scores in all three of its rounds, which had occurred only once before in World Festival history. Jonathan wrote Chapter 1, "The LEGO MINDSTORMS NXT System," Chapter 15, "ScanBot: An Image-Scanning Robot," and Appendix A, "Differences Between Sets." He also wrote part of Chapter 5, "Making Sense of Sensors."

Brian L. Davis

Dr. Brian L. Davis has been building LEGO robots for about five years, competing in conventional events such as sumo, line following, and maze solving, as well as developing other more advanced robots, such as automated forklifts. He has also worked on community projects such as the Great Ball Contraption (a LEGO cooperative kinetic sculpture) and coordinated at various fan events such as BrickFest, Brickworld, and House of Bricks. In 2005, Brian was invited to be a member of the MINDSTORMS User Panel expansion, and since then he has immersed himself in the NXT, especially the NXT-G language, developing novel robots and solutions. He received a PhD in physics from the University of Michigan and currently divides his non-LEGO time between being a stay-at-home father of three and teaching college-level physics, biophysics, and astronomy.

Brian's contributions include Chapter 2, "The Grammar of NXT-G," Chapter 3, "NXT-G Problems and Solutions," Chapter 7, "Bluetooth on the NXT," and Chapter 8, "NXT-to-NXT Remote Control." Because he just couldn't stop, he also co-wrote Chapter 9, "RaSPy: A Rock, Scissors, Paper–Playing Robot."

Jim Kelly

Jim Kelly is a freelance writer based in Atlanta, Georgia. He received an English degree and an industrial engineering degree—an unusual combination, but very helpful in his career. Jim was accepted into the MINDSTORMS Developer Program (MDP) in early 2006 and helped to beta test the LEGO MINDSTORMS NXT kit and software. He is now a member of the MINDSTORMS Community Partners (MCP), a group that continues to assist LEGO with testing and growing the NXT product. Jim cannot wait to introduce his newborn son to robotics, and his wife is looking forward to another set of hands to help clean up all the LEGO pieces on the floor. Jim's contributions to the book include Chapter 4, "Debugging—When the Unexpected Occurs," and Chapter 11, "3D PhotoBot: A 3D Photo Assistant Robot." He also worked with Brian Davis and Matthias Paul Scholz on Chapter 9, "RaSPy: A Rock, Scissors, Paper–Playing Robot."

David Levy

David Levy is the founder of RestonRobotics.org, a network that enables students, parents, and teachers to collaborate on robotics projects. He is also actively involved with outreach as an education director with the Virginia FIRST LEGO League. He became a contributor to The NXT STEP blog in early 2006 and since then has revamped the site to offer forums, videos, and book discussions. David resides in northern Virginia with his wife and three children, where he is currently employed as a software architect for a local startup. David is responsible for the book's official website, as well as video content and tutorials that readers can download and use to further their explorations with the book's robots and ideas.

Lou Morris

Based in Ottawa, Canada, Lou Morris got his first taste of computers, programming, and automation in the early 1980s. Being self-employed from an early age has given him the time required to research, develop, and play with leading-edge electronics and robotics. Recently, Lou has headed several software development projects for various technologies, including a treasure-hunting underwater remotely operated vehicle (ROV) robot; a medical device used to treat cerebral palsy patients; and the roboDNA PC Dashboard Designer for the LEGO NXT, iRobot, and PIC microcontroller. As a contributor to The NXT STEP blog, he helps to cultivate the LEGO MINDSTORMS NXT community. Lou is responsible for Chapter 6, "Design."

Fay Rhodes

Fay Rhodes is a freelance designer who provides desktop publishing and web design services to nonprofit organizations in the Boston area. She designs NXT robots for publication and is experienced in the use of CAD software for LEGO. Fay is tech editor of *LEGO MINDSTORMS NXT-G Programming Guide* (Apress, 2007). Thanks to her husband, Rick, she has the clearest and most accurate building instructions available. Fay co-wrote Chapter 10, "Beach Buggy Chair: A Ramblin' Robot," with her husband, Rick, and co-wrote Chapter 5, "Making Sense of Sensors," with Jonathan Daudelin and Matthias Paul Scholz.

Rick Rhodes

Rick Rhodes is a frequent contributor to The NXT STEP blog from Framingham, Massachusetts. He develops robots that appeal to young teens because he has a teenage son who enjoys the NXT. Rick also critiques the NXT creations of his wife and fellow contributor, Fay Rhodes. Words cannot describe how deeply she appreciates this. Rick worked with Fay on Chapter 10, "Beach Buggy Chair: A Ramblin' Robot."

Matthias Paul Scholz

Matthias Paul Scholz (from Freiburg, Germany) has a degree in mathematics and has held IT-related positions in various companies in Germany over the past 12 years. He has been an active member of the LEGO MINDSTORMS community since 2000, was one of the developers of the open source leJOS platform for the RCX, took part in the MINDSTORMS Developer Program (MDP), and is presently one of the 20 members of the MINDSTORMS Community Partners (MCP) program. Furthermore, he wrote *Advanced NXT: The Da Vinci Inventions Book* (Apress, 2007), contributes to The NXT STEP blog, and maintains the German-language sister blog Die NXTe Ebene. Matthias contributed Chapter 13, "Slot Machine: A One-Armed Robot," and worked with Brian Davis and Jim Kelly on Chapter 9, "RaSPy: A Rock, Scissors, Paper–Playing Robot."

Christopher R. Smith

Christopher R. Smith (a.k.a. Littlehorn) is a senior quality assurance inspector in the Shuttle Avionics Integration Laboratory (SAIL) at NASA's Johnson Space Center in Houston, Texas. He invented an inspection tool recognized by NASA, which honored him with the prestigious Space Act Award. He has been designing LEGO MINDSTORMS robots since rediscovering the LEGO product in 1997. As one of the pioneering moderators asked to host the official LEGO MINDSTORMS website community forums, Chris has volunteered there for the last nine years and helped to cultivate one of the safest online communities. He was again honored when LEGO asked him to become a member of the MINDSTORMS Developer Program (MDP) for the NXT system. He is a member of the MINDSTORMS Community Partners (MCP) and is a contributor to The NXT STEP blog. Chris wants to thank his wife, Veena, and his children, Revi and Benjamin, for the inspiration and the motivation to *LEg GOdt* (*play well*). Chris wrote Chapter 14, "BenderBot: An Anti-Theory Music Robot," and Appendix B, "Trouble-Free CAD Installation Guide."

Rob Torok

When he is not busy with his wonderful family (Anita, Mitch, and Rohan), Rob Torok is a teacher of mathematics, information systems, and robotics (of course!) at Claremont College, a senior secondary school in Tasmania, Australia. In addition, he has an online robotics class of more than 70 students in about 20 schools across the state. Rob is also the chair of Robotics Tasmania, the organization that conducts the annual RoboCup Junior Tasmania competition for school students. Prior to the official release of the NXT, he was a member of the MINDSTORMS Developer Program (MDP). Rob's contribution to the book is the humorous "Marty: A Performance Art Robot" found in Chapter 16.

brief contents

contents in detail

PART II THE ROBOTS

foreword: why LEGO matters

Let me tell you why I care about LEGO. It started, naturally enough, with my kids. I've got a lot of them—four, with a fifth on the way—and they're all pretty young. Like all dads, I'd like to spend more time with them, but I've got the whole busy job thing. Plus, and let's be honest here, I'd like a little more time to do things that are fun for me, too. Pokémon cards just don't do it.

Fortunately, what both my kids and I like best is to build things. We want to invent, to create whatever our imaginations can come up with, to make something *real*. Given that metal lathes and machine tools are a bit out of our range, we did what millions of people had done before us: We fell in love with LEGO. For my kids, it's a toy. For me, it's a prototyping tool. But the common factor is that all of us look at a box of LEGO pieces and we see *infinite potential*. To mangle a phrase coined by Microsoft, "What will you build today?"

For the first few years of our love affair with LEGO, we did the usual: built models according to the instructions, and then rebuilt them according to our own whimsy. Then, thanks to the Web, we started looking at what other people were doing with LEGO and were, like everyone who discovers the LEGO subculture, blown away. The incredible creativity, ambition, and technical skills! (Have you seen that LEGO Rubik's Cube solver?) And all channeled through something that's sold as a toy. There was clearly something extraordinary happening here.

During the build-up to the MINDSTORMS NXT release, I was fortunate as the editor-in-chief of *Wired* magazine to be in touch with LEGO company officials. They gave us an inside look at how the company was working with adult LEGO fans to develop a product that could harness both the energy of amateurs and the skills of professionals to make something better than either could create alone. We profiled this amazing "pro-am" partnership in a cover story in *Wired*, headlined "The LEGO Army Wants You." (One of my kids' minifigs, cloned many times in Photoshop, was the cover image.)

We were hooked. We had an early version of the MINDSTORMS NXT kit and we proceeded to build. First we built the recommended models. Then we tweaked them. Then a friend showed us how to make a LEGO photocopier. Then we started thinking about what we could make that had *never been done before*. Here's what we did. We'd been playing with some radio controlled model planes that had shown up in the *Wired* offices for review. We'd done the usual crashing and rebuilding and, while fun enough, we didn't think it was hugely creative. But the combination of LEGO and R/C planes laid the ground for our next level of obsession.

One evening, while playing with TECHNIC and MINDSTORMS parts with the kids, I was browsing the LEGO blogs when I read that a company called HiTechnic was releasing a MINDSTORMS Gyro Sensor. They demonstrated it with a video of a LEGO Segway.

Holy cow. I sat there with my jaw on the floor for a minute, then decided to go for a run to do a bit of contemplating. On the road, I considered the implications of a MINDSTORMS Gyro Sensor.

Suddenly it came to me: I needed to create a LEGO autopilot. A gyro-enabled NXT brick that could fly one of our R/C planes. I raced home, built a prototype on the dining room table, and uploaded a picture and description to my blog that evening. The next morning I awoke to find that my post had been picked up by some of the biggest tech news sites around, including Slashdot and Digg, and I'd had nearly a hundred thousand page views.

The rest is history. Over the past year, my kids (well, the older ones, at least) and I have turned that LEGO autopilot idea into a real unmanned aerial vehicle (UAV) that uses a MINDSTORMS NXT brick to fly a plane, all by itself. The Gyro Sensor turned out not to be the solution (for complicated reasons, it's too hard to get absolute position from that kind of inexpensive "rate gyro"), but we found ways to create an autopilot nonetheless. We developed our prototype using a Compass Sensor, and then worked with one of the "LEGO Army," Ralph Hempel, to integrate a cheap commercial GPS sensor via the NXT's built-in Bluetooth connection. (You can follow our progress and build your own UAV with our instructions at http://diydrones.com.) Our next challenge is to do the same with a Bluetooth mobile phone, so that we can send the plane commands via text message.

After that, who knows what we (or you!) will do. With LEGO, even the sky's not the limit.

Which brings me to *The LEGO MINDSTORMS NXT Idea Book*. I'm a regular reader of The NXT STEP blog (http://thenxtstep.com), the creative home of the contributors to this book. I find the blog to be a terrific source of ideas about building with the NXT kit, and reading it has become almost an addiction. Fortunately for all involved (both you and I as readers), the contributors to The NXT STEP blog have done a superb job of bringing their enthusiasm and ideas together to create this very exciting book. You'll find in-depth treatments of programming, design, Bluetooth, and troubleshooting, as well as very LEGO-like plans for building a number of robots. I recommend this book to you, and I hope that you'll enjoy reading it as much as I have.

Chris Anderson
Editor-in-chief, *Wired* magazine

introduction

Welcome to *The LEGO MINDSTORMS NXT Idea Book*! As you can tell from the title, this book is all about the MINDSTORMS NXT robotics kit. But there really is so much more to the story, so before you dive in, we hope you'll read on and discover a little bit more about how this book came to be, as well as the people responsible for its creation.

The NXT STEP blog

In January 2006, a group of us started a blog called The NXT STEP to cover news and information about the new NXT robotics kit. At the time, the kit was an early beta and not available to the public. The kit wasn't due to be released until August 2006, but pictures and bits of news were being slowly released by LEGO, and the MINDSTORMS fan base was anxious to know more. As we write this, the readership of The NXT STEP has grown to more than 35,000 readers from more than 100 countries, with more than 18 contributors adding to the site.

In late 2006, three of the early contributors to The NXT STEP began to work on a robot together, even though they were spread across the globe (two in the USA and one in Germany). (You can read more about this early collaboration in Chapter 9.)

It was this concept of collaboration in robot building and programming that led to the book you are holding. The NXT STEP team began to discuss sharing its knowledge of robot building and programming (as well as some other concepts), and the idea for a book was formed. The book would contain not just instructions for building robots but also chapters on related topics such as programming, debugging, building, and more. Our goal was to give you, the reader, the information you need to begin investigating the power and flexibility of the NXT.

Contributors began to submit ideas for chapters, while helping one another by testing robots, debugging programs, and providing additional information. The book grew in size, as well as diversity, as contributors displayed their own building and programming styles.

who is this book for?

We made certain assumptions when writing this book. First, we assume that you either own a LEGO MINDSTORMS NXT Retail version or have access to one. While some of the robots can be built with the Education version (sold by the LEGO Education division), there are certain parts that are not included in the Education version. If you own the Education version, we recommend that you purchase the Education Resource Set, which includes additional parts that will be useful for completing these robots. Appendix A contains a more detailed breakdown of the two versions of the NXT kit. You should read it carefully if you are considering purchasing a kit and are unsure of the differences.

We also assume that you have some familiarity with building and programming with the NXT. The basic NXT kit includes several tutorials, and we encourage you to follow along with these tutorials to gain a better understanding of how to build and program robots with the NXT.

You should be able to easily duplicate the robots that you find in this book by following the building instructions provided. Programs with detailed descriptions of settings for each software component are also provided, all of which can be duplicated using either the Retail or Education versions of the software.

about this book

Our book is broken into two parts. Part I, "Beyond the Basics," contains chapters that will help you better understand some of the topics related to working with NXT robots. Chapter 1, "The LEGO MINDSTORMS NXT System," offers a general overview of the components of the NXT system, including a comparison with the earlier MINDSTORMS RIS system. Chapter 2, "The Grammar of NXT-G," discusses how NXT robots receive instructions. Chapter 3, "NXT-G Problems and Solutions," and Chapter 4, "Debugging—When the Unexpected Occurs," help with troubleshooting and provide debugging tips that help to you solve problems and make your programs work more efficiently. Chapter 5, "Making Sense of Sensors," digs into sensors. Chapter 6, "Design," is a nice tutorial on building technique with some great tips for increasing your building skills. Chapter 7, "Bluetooth on the NXT," tells you what you need to know to get your NXT brick talking with other Bluetooth-enabled devices; and Chapter 8, "NXT-to-NXT Remote Control," will help you to remotely control your robot.

Part II, "The Robots," contains the actual robot building and programming instructions. The robots include the following:

* RaSPy: A Rock, Scissors, Paper–Playing Robot plays a game.
* Beach Buggy Chair: A Ramblin' Robot is easy to build, fun to use, and features a unique wheel arrangement.
* 3D PhotoBot: A 3D Photo Assistant Robot will help you take photographs that can be converted into three-dimensional images.
* CraneBot: A Grabber Robot is an advanced robot that demonstrates many unique building techniques.

* Slot Machine: A One-Armed Robot will feed your gambling addiction (and take your money).
* BenderBot: An Anti-Theory Music Robot will have you exploring your creativity in the new world of musical anti-theory.
* ScanBot: An Image-Scanning Robot scans black-and-white pictures and displays these images on the NXT's LCD.
* Marty: A Performance Art Robot will help you create some random and interesting art.

The book's appendixes include discussion of the two versions of the NXT kit as well as details on installing and using the software that was used to create the building instructions in Part II. Appendix A compares the different NXT sets, and Appendix B offers a guide to installing and updating LEGO computer-aided design (CAD) programs.

how to use this book

The best way to use this book is to scan through it and flag the sections that you find most interesting. If you're familiar with NXT and are ready to start building a robot, that's the best place to start. If you're a little nervous or are unfamiliar with NXT, start with Chapter 1; then continue with Chapter 2. After that, feel free to skip around. But above all, just enjoy the book and have fun. We had fun putting it together and we hope you have just as much fun following along!

PART I:

beyond
the
basics

the LEGO MINDSTORMS NXT system

Robots are fascinating. The concept of nonliving materials performing complicated tasks all by themselves is truly amazing! Since robots are so complicated and high-tech, you'd think it would require a lot of knowledge and skill to design and program them. But in fact, LEGO MINDSTORMS makes robotics easy and fun for children and adults alike.

MINDSTORMS robotics started back in 1998 with the Robotics Invention System (RIS), sold by LEGO. Parts such as motors, sensors, and microcomputers could be combined with other ordinary LEGO construction pieces to make working robots. The RIS also provided easy-to-use programming languages that enabled you to program your robots to act on their own.

Now, with the LEGO MINDSTORMS NXT, users have been given many improvements over the RIS, making it even easier and more enjoyable to create robots.

The NXT kit goes way beyond just improvements in the software and hardware, however. The new kit opens up robotics to all ages and takes away the complexity and "techy" feel of other robotics kits. But changes are what make the two kits different, so keep reading for some more details on what makes this new kit so friendly and such a welcome upgrade.

> I most appreciate the NXT for the interesting (and brain-stimulating) creative possibilities it offers. NXT is more entertaining than puzzle books, and, as a bonus, there is a significant "cool" factor with the children and grandchildren.
> —Fay Rhodes

differences between the RIS and the NXT

Although the NXT system was designed to improve on the RIS, as is often the case with any new product, many fans of the old system were initially dissatisfied with some of the changes. One concern was with the new type of construction pieces in the NXT system (which are studless). In the RIS, most of the construction pieces are like regular LEGO building blocks, in that they have *studs* (those small round parts of bricks that stick out on top, giving a classic "LEGO" look to them). The NXT system's construction pieces, however, are almost all TECHNIC pieces, which have no studs. (As a matter of fact, you'll hear many users talking about "studless building," and these new TECHNIC pieces are exactly what they are talking about!) Once you're used to building with studs, it can be difficult to get used to studless building. However, we think you'll find that studless construction actually makes it easier to build stronger and more flexible designs.

Building with TECHNIC pieces truly does improve your robot designs; pieces connect together with a stronger hold and because of the variety of shapes of pieces, the new kit offers up many new variations on robot shapes. With the RIS, many users complained that robots always looked "boxy," and square-shaped robots were the norm because they were built with standard LEGO bricks (which are rectangular and stackable). NXT robots, on the other hand, can take on a wider variety of shapes (as you'll see with the robots in this book), and users actually enjoy trying to build unique designs that haven't been seen before. And NXT robots look more, well, robotic.

Another complaint was the fragile nature of RIS robots. Many RIS robots suffered a harsh fate when falling just a few feet, shattering into dozens or even hundreds of pieces. No more! NXT TECHNIC pieces are more rugged and stay together; they'll still come apart if a robot is dropped from a sufficient height, but the pieces connect together better. All in all, we think you'll find working with the TECHNIC pieces rewarding and easy to understand; you'll be snapping pieces together and coming up with new designs in no time.

You might also hear people expressing concern about the increase in size of the NXT's electrical parts. The motors and sensors in the NXT system are significantly larger than those in the RIS. Although this larger size can make it harder to fit the motors and sensors into your designs efficiently, many of these parts have improved features that balance out this disadvantage. For example, although the NXT motors are much bigger than the RIS motors, they include built-in Rotation Sensors—a huge advantage when you want to build robots that move precisely.

Yes, you'll have to figure out how to properly connect these odd-looking motors to your robots, but that's half the fun, anyway. Throughout this book you'll find examples of robots that use these motors in unique ways, which will provide you with ideas that you can use in your own robots. Don't let the motors' size bother you; take the increased power and the ability to control motor movement (in both degrees and rotations) and put them to work!

Perhaps one of the least-liked changes is the addition of $50 to the price of the base set. The RIS base set cost $200, while the NXT set has a price tag of $250. However, the extra money spent on the NXT set gets you all the advantages of the new system, so most people think it's worth it.

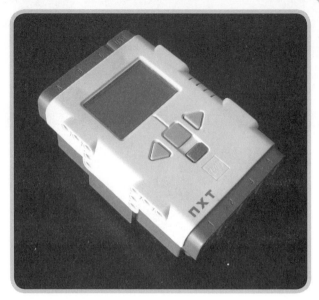

Figure 1-1: The NXT brick

> What I like best about the NXT (and the MINDSTORMS range in general) is that it provides such an easy way into robotics. I love being able to turn an idea for a robot into something that actually works without having to go near a soldering iron.
> —Rob Torok

the electronics

LEGO sells a base set that contains all the basic features of the NXT system. It includes several electrical parts—among which are a microcomputer, sensors, and motors. The microcomputer is called the *NXT brick*, and it's an intelligent, computer-controlled brick that acts as the brain of your robotic designs. Programs direct it to receive input from sensors, activate motors, play sounds, and more.

The NXT brick (shown in Figure 1-1) has seven main features, two of which have to do with letting you download programs to the brick. On one end of the brick is a port to which you can connect a USB cable. After the cable is connected, you can use it to download programs to the brick. The brick also has Bluetooth capabilities that enable you to wirelessly download programs and communicate with other bricks, Bluetooth-enabled phones, and other Bluetooth devices.

An LCD on top of the brick can display images, text, and drawings; and a speaker can play tones (as the RIS could), as well as prerecorded sound files. For example, you could program your robots to say phrases such as "Hi!" and "Good job!" through the speaker. This feature enables you to give your robots a new level of animation and make them a lot more fun.

To power motors and receive input from sensors, the NXT brick has three output ports and four input ports. Sensors can be attached to the input ports, which are numbered 1 through 4, by connector cables included in the NXT system. Once connected, the sensors send information about the environment to the brick, which can then be used to affect the robot's behavior. Motors can be attached to the three output ports—A, B, and C—which can then be used to make the robot walk, pick up objects, or perform many other movements.

> What do I like about the NXT? Bluetooth. The ability to control my robot using my cell phone or PDA with free software from the NXT community. Sometimes I build a robot and just want to see it in action without spending a lot of time programming. With Bluetooth, I can quickly make a connection and send commands to the motors to test my design.
> —Jim Kelly

The NXT's motors (shown in Figure 1-2) are servo motors. They are more powerful than the RIS motors, thus enabling you to build stronger and faster robots. They also have built-in Rotation Sensors, which measure the motors' revolutions (in rotations or degrees), a feature that enables your robot to make very precise movements.

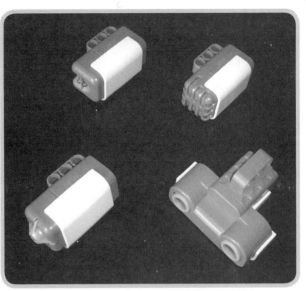

Figure 1-3: The Touch Sensor, Sound Sensor, Light Sensor, and Ultrasonic Sensor

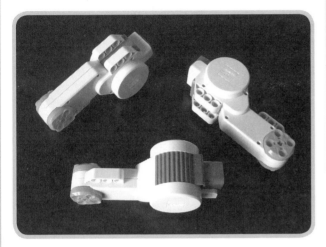

Figure 1-2: NXT servo motors

There are four types of sensors in the NXT system, and they measure touch, sound, light, and ultrasound (see Figure 1-3):

* *Touch Sensors* have a button that senses when it is pressed, released, or bumped. This sensor is useful in robots that need to detect obstacles or react to touch.
* *Sound Sensors*, new additions to the NXT system, monitor sound in the environment. Robots can use these sensors to react to voice commands.
* *Light Sensors* detect the intensity of light around them, and they are also equipped with an LED light source so that your robot can determine the intensity of reflected light. Light Sensors enable your robot to do things such as sort bricks by color or follow a line.
* The *Ultrasonic Sensor* is new to the NXT system. It uses reflected, ultrasonic sound (sound we can't hear) to measure the distance between the sensor and an object. This sensor has many uses, such as mapping a robot's environment, detecting objects without hitting them, or detecting motion.

Currently there are also Compass Sensors, Acceleration Sensors, Color Sensors, and Temperature Sensors, and more sensors are being released as you read this. LEGO and third-party companies such as HiTechnic (http://www.hitechnic.com) are dedicated to increasing the number of sensors that will work with the NXT, and you'll find that these sensors give your robots more functionality.

Are you wondering about new electronics? LEGO will be releasing some new motors (they might actually be out when this book hits the shelves), and expansion packs are expected in the near future. LEGO is aware of the popularity of the new NXT kit, and you can be assured that it is working hard to provide new and improved features and components to users who are hungry for bigger, faster, smarter, and more complex robots.

The introduction of the servo motor and its NXT programming interface is the main reason why the NXT is accessible to an even younger audience than its predecessor. When using the RIS, in order to ensure accuracy in traveling distance, one had to attach one or more sensors to measure the rotations of the wheel motors. However, with the NXT, two synchronized wheel motors can be controlled with a single Move block to enable the robot to travel in a straight line or turn at an angle. That's a much simpler way to control movement.
—David Levy

the building pieces

To build the framework of the robot, the NXT system has construction pieces, as you'd expect from LEGO. They aren't the typical LEGO bricks that we usually think of, however; they're studless. As mentioned earlier, the NXT system construction pieces are TECHNIC pieces. Although you might find that it takes a while to get used to building with these studless pieces, we have found that they allow for more flexibility and strength in our robot designs.

Along with the basic TECHNIC pieces such as beams, pins, and axles, the NXT Base Set includes several other pieces that the RIS didn't include. For example, the NXT Base Set includes two LEGO balls, a turntable, and claws. Many of these pieces were added to make it easier to build designs with TECHNIC pieces only, and some were added simply to open up more robotic possibilities. Figure 1-4 shows some of the pieces in the NXT that weren't in the RIS.

Figure 1-4: New pieces in the NXT system

As it turns out, the NXT Base Set even introduced a new LEGO construction piece. This piece, nicknamed the *Hassenpin* (after one of the individuals responsible for pushing LEGO to create it), is very useful for making right angles in TECHNIC designs without using bricks. The retail NXT Base Set includes eight of these handy little pieces (and you'll be wanting more).

All in all, you'll find that the variety of components included in your kit will provide you with an almost endless supply of robot designs. With 519 components, you shouldn't find yourself running out of pieces (or ideas!) any time soon.

the programming language

Although the previously mentioned parts make up the entire physical design of the robot, they won't do anything without a program. The NXT system's programming language, which is called *NXT-G*, is a graphical, drag-and-drop language that we have found to be both very easy to use and powerful. If you've used the ROBOLAB software with the RIS, you'll probably see some similarities.

When programming, you can choose from many different types of programming blocks, each of which directs the robot to perform a specific action (such as moving a motor, playing a sound, and so on). By stringing several of these blocks together, you can create a list of actions for the robot to perform, exactly as you want.

You'll find blocks that receive readings from sensors and blocks that create random numbers and sounds. Some blocks enable you to put text on the LCD, while others give your NXT robots the capability to do some very complex mathematical calculations.

One of the biggest surprises for LEGO MINDSTORMS fans was the ability to create reusable blocks called My Blocks (shown in Figure 1-5). *My Blocks* enable the programmer to reuse certain types of programming sequences (such as a series of steps that would have the robot move forward 3 rotations and then turn left 90 degrees). Grab this series of steps as a My Block, drop it into your new program, and save yourself some steps. (For a more thorough introduction to My Blocks, see "My Blocks Save Time and Simplify Your Programs" on page 16.)

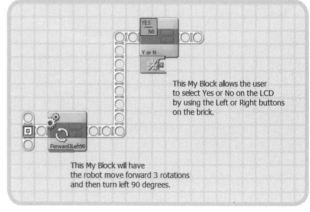

This My Block allows the user to select Yes or No on the LCD by using the Left or Right buttons on the brick.

This My Block will have the robot move forward 3 rotations and then turn left 90 degrees.

Figure 1-5: My Blocks

NOTE My Blocks can be shared with other NXT-G programmers, and libraries are already popping up on the Internet in which NXT fans are sharing their custom My Blocks.

The biggest surprise with NXT-G, however, is that it is so easy to use! Users with absolutely no programming experience can be up and running in no time. LEGO wisely chose to include a variety of programming tutorials with the software; these demonstrate many of the basic programming blocks, as well as some programming techniques that will benefit beginners and advanced users alike. The graphical user interface (GUI) is so easy to use and intuitive that many people just jump right in and begin to experiment with the software, figuring out how it works by trial and error.

The NXT system's flexibility also enables it to be programmed by using other languages. Three of the most common third-party languages are NBC, NXC, and RobotC. NBC and NXC are free languages created by John Hansen. They are both text-based languages, and NXC is similar to the C language (*NXC* stands for *Not eXactly C*). They can be downloaded from http://bricxcc.sourceforge .net/nbc. RobotC is also a text-based language that is a lot like C. A product of Carnegie Mellon University's Robotics Academy, it can be purchased from http://www.robotc.net.

We encourage all readers to go through all the tutorials included with the NXT software. You'll learn how to place blocks on the screen, configure them to do all kinds of interesting things, and then upload that group of blocks (called a *program*) to your robot, which is ready and willing to take those blocks and do something useful with them.

conclusion

All in all, the NXT system is a simple, fun, yet powerful robotics system. It provides microcomputers, sensors, motors, and building parts that can easily be connected and programmed to make robots move and work by themselves.

Of course, one question we get a lot is, "Where do I start?" And this book is one answer (it is called the *Idea Book* for a reason). We have tried to include a variety of robots in this book, not only to demonstrate some building techniques with the new TECHNIC pieces but also to show you the power of an NXT-G program.

This chapter discussed the new kit and gave you a great overview of the pieces and parts, but until you roll up your sleeves and start to build and program, you haven't really experienced the NXT. Keep reading to find out just how the NXT will become your new favorite toy . . . um, tool.

A BRIEF ARTICLE ON ARTICLES

There seems to be some debate as to whether a single NXT should be called *a NXT* or *an NXT*. The debate stems from the fact that in English, which article you use depends on the first sound of the following word . . . but there are two different ways that *NXT* can be pronounced. Pronounced as a "word" it starts with a consonant sound (a "next"), but if you pronounce each letter, it starts with a vowel sound (an "en-ex-tee"). LEGO seems to have wisely avoided the entire issue, not using either article in its text, but instead words like *the*, *your*, *other*, and so on. Since neither way is "the right way," it's your choice whether you use *a* or *an* (or both!), and you'll see it both ways in this book.

the grammar of NXT-G

The first problem most people have with the LEGO NXT MINDSTORMS kit is the studless architecture. Where did all the studs go? How can you connect the pieces firmly, in the ways you want to? With a little bit of help from the building instructions and some experimenting, you can cross that hurdle only to immediately face another one: How do you program your creation? What are all these funny icons snapped to a beam on the screen, and how do they work?

is NXT-G a toy programming language?

For some people used to a text-based language, NXT-G just looks . . . well, weird. Instead of a series of ordered instructions, it looks as if we have a bunch of brightly colored blocks lined up along the midline of the screen. Even more confusing, few text clues onscreen hint at what these various blocks are trying to do or how they work together. As a result, a lot of folks are likely to throw their hands up in frustration, decrying the fact that LEGO and National Instruments (NI) created a "toy language."

Well, to a point, they're correct.

NXT-G is indeed a computer language designed to explicitly run on a toy; after all, that is what the NXT is and what LEGO produces. So any language that ships with the NXT should be easily used by the target audience, which is kids . . . *not* a bunch of folks with years (or decades!) of experience programming in high- and low-level text-based languages.

But don't let that fool you into thinking that NXT-G is "just a toy" or that it can't be used to write "real programs." True, some things have been limited or hidden to make the environment a little easier for a first-time user to navigate (and true, there are some things that are missing), but there is a lot of power under the hood in NXT-G. The trick is that in order to use it, you have to approach it as a *new language* and learn its intricacies and strengths, as well as its weaknesses.

NXT-G is no more a crippled toy than the graphical user interface (GUI) on your PC at work is a toy—humans can use pictures to communicate a lot of information rapidly, and graphical languages (and interfaces) take advantage of this.

NOTE NXT-G was developed on top of (and owes a lot to) a highly regarded professional development language called LabVIEW. Far from being a toy, LabVIEW is used in complex data acquisition and control systems all over the world, serving as a flexible and powerful tool for scientists and engineers. Since a full LabVIEW environment retails for hundreds to thousands of dollars, you can consider NXT-G a wonderful and inexpensive way to sample a graphical programming language.

Now I'll be the first to admit that there's no way to explain the entire NXT-G language in a single chapter. There's just too much to talk about. There are 28 stock blocks (and a lot more can be added), and each block can have as many as 8 different sections in its configuration pane, as well as multiple configuration panes, and so on. In fact, an entire book could be written about programming in NXT-G. Instead, I'm going to try to explain some of the quirks and tricks that you can use to do much more in NXT-G. I'll assume that you have programmed at least some in NXT-G, but need a little boost to understand some of the oddities about the language or how to use some specific blocks. But this really begs the question of how you get this basic competency.

where's the guidebook?

The first step most people take in learning NXT-G is to work through at least some of the programming examples in the Academy content (those detailed "build and program" instructions that are initially available along the right-hand side of the main window). Figure 2-1 shows the Academy area.

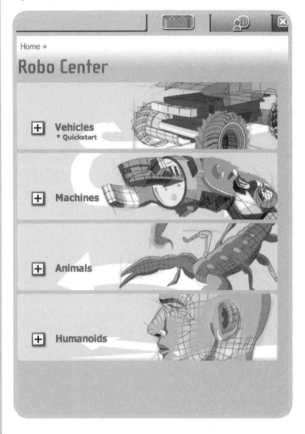

Figure 2-1: The Academy area as it appears when the program starts

While these instructions will walk you through building a program step-by-step, they don't offer much at all in the way of detailed explanations. So how can you learn the language? Where did LEGO put the programming guide?

help is built in

LEGO put the programming guide right into the programming environment: Just hover (don't click) the mouse pointer over any block

or structure on the programming sheet, and a brief explanatory note will pop up in the Help pane in the lower-right corner of the window. Actually, there are two useful panes down there: Clicking the upper question mark tab brings the Help pane forward, while clicking the lower magnifying glass tab brings the navigation pane forward. More about that later.

These brief notes are helpful, but the real secret is what's under them. If you click the *More Help* link, NXT-G obediently opens a full, detailed, linked web page of information about almost every aspect of the block, including what each input or output plug does, what the limits on certain parameters are, how it can be used (and in some cases how it shouldn't be used), and so on.

help index

You can also open this detailed help from the NXT-G menus by selecting Contents and Index from under the Help menu in the menu bar. The index that opens along the left side of your screen is valuable (see Figure 2-2). It not only contains the help files for every NXT-G block, but it also starts with some important overview subjects, such as sequence beams and data wires, for example. In short, LEGO and NI have included a wonderful, detailed, block-by-block description of the language right in the product.

For example, open up the Help window and click *Starting Point* in the General Topics section in the left pane. This not only tells you that the odd little symbol at the start of your program has a name, but lets you know that by selecting it, it opens up a comments text box in the configuration pane.

You can document anything you like about your program here: your name, when you wrote it, special hardware it needs, your grocery list, and so on.

There are many more little gems like this in the documentation, but you have to actually *read* the documentation to find them. (If I sound like your fourth-grade teacher, there's a reason; teachers know that a student usually has to actively process something to learn it, not just have it available and unread.) How many times have you heard the words *Did you read the instructions?* Well, in support of beleaguered spouses everywhere, I have to repeat them: Read the instructions (now that you know where they are).

the basics: starting out

First, let's lay out some terminology—not because terminology is fun, but because it's useful to have a common language in which to talk (and think) about such things.

The hard-copy user's manual that comes with NXT does a great job of describing the programming environment, including things like the configuration panel (or configuration pane), the

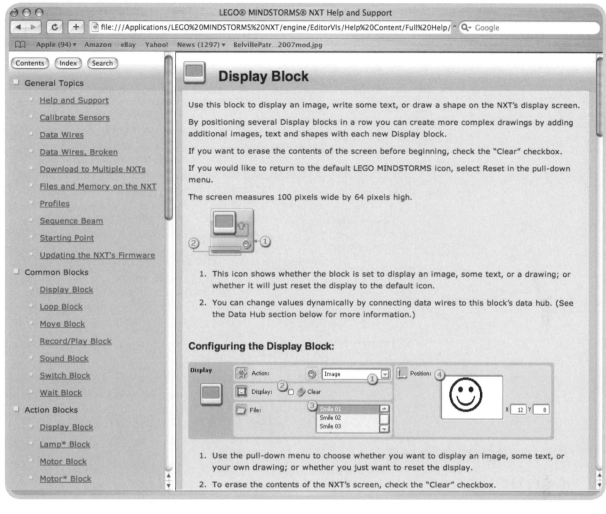

Figure 2-2: The Help HTML screen with the index and the page describing the Display block

work area (sometimes called the *programming sheet* because you can have multiple sheets open and hidden behind each other like sheets of paper), and so on. But as far as elements of the actual NXT-G programs, the manual doesn't go into great detail.

There are three basic elements used to build programs in NXT-G: sequence beams, blocks and structures, and wires. Okay, that's four, but blocks and structures are really very closely related. Figure 2-3 shows each of these elements.

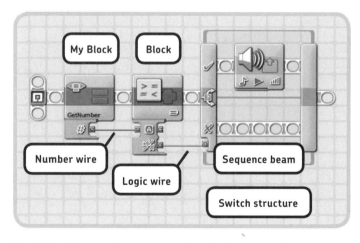

Figure 2-3: A simple program that shows sequence beams, blocks, wires, and a Switch structure

sequence beams

Sequence beams are the white studless beams that link blocks together onscreen. The order of blocks along the sequence beam controls the order in which they execute: The first block connected to the start symbol executes first; then (usually when the first block is completely finished) the next block along that sequence beam executes; then the third one, and so on. Sequence beams are just a visual way to represent the program flow, not unlike the arrows in a flowchart, as you can see in Figure 2-4.

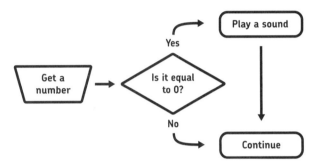

Figure 2-4: A simple flowchart for the program shown in Figure 2-3

branching and sequences

Like a flowchart, these "execution sequences" can branch, but they branch in two very different ways. The first is when a choice has to be made: Should the execution follow this path, or that one? In a text language this is done with something like an if-then or a switch-case command, while in NXT-G the Switch block takes over this role (more on that in a second).

And NXT-G offers another very different, very powerful way to make the execution proceed: You could have it go down two different branches simultaneously. This is *parallel programming*: using two or more parallel sequences to accomplish two things at once.

In NXT-G, parallel programming is as natural as breathing. Once you have laid down one sequence (or even a portion of that sequence), you can start a second sequence by dropping a block somewhere below it, *not* connected to the existing sequence beam

(actually, it could be below it, above it, or anywhere else, as long as it doesn't link up to the sequence beam). Such an "orphan" block is normally ignored by NXT-G and not incorporated into your finished program.

Now, to branch a sequence beam and link in this first block of a parallel sequence, hover the arrow cursor over the existing sequence beam at the point you want it to branch, and press and hold the SHIFT key. You can now click and drag out a new branch of the sequence beam to the start of your orphan block, and that's it—you've started a new parallel sequence. Any blocks you drop behind this newly connected block will obediently join the rank and file of the second sequence. You can branch a new sequence from anywhere: the start symbol, midway along a sequence beam, or even within a Loop or Switch (although that's a bit trickier). Figure 2-5 shows a sample three-pane sequence.

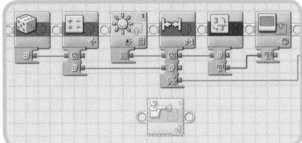

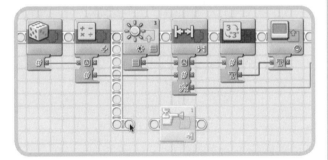

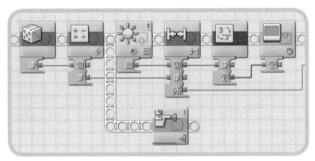

Figure 2-5: A three-pane sequence of creating a parallel sequence beam

blocks and structures

Blocks and structures are the real workhorses of NXT-G. They are specialized chunks of code that do something specific, such as play a sound, control a motor, get the current reading from a sensor, and so on.

Almost any time you place a block or a structure, you need to set up its various limits and conditions from the configuration panel, such as what port a sensor is connected to, how long to run a motor, and so on—and there can be a lot of things to configure.

blocks

Of the two, *blocks* are the simplest; they're just one-square icons that accomplish something specific. Program execution usually halts while a block is executing; that is, the next block on the sequence doesn't start doing its thing until the previous "upstream" block has completely finished. But that's not always the case: Some blocks can start actions and then let the sequence continue executing, even while these actions are still taking place. A good example of such a block is the Sound block, which enables you to check or uncheck the box next to the words *Wait for completion*.

cloning blocks

Re-editing the configuration panel every time you drop a new block gets old fast. Every block has only one default state when dropped from the palette, but that's not much help if you want to repeat a block that is set with something other than the default state.

To repeat an existing block you can clone it. If you ALT-drag an existing block (OPTION-drag on the Mac), the environment makes a "clone" of the block you selected—copying every setting from its configuration pane into the new one, and even imitating the data plugs that are currently displayed. This is another wonderful feature inherited from NXT-G's "parent language" LabVIEW, and it's great that the developers left such things in to help the advanced user.

loop and switch structures

The Loop and Switch structures function almost like composite blocks, or blocks that contain other blocks. When they are first placed they do nothing, because there is nothing inside them to do; you must drop blocks into them to set up their sequences.

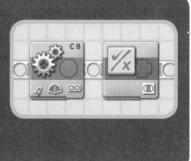

Any sequence within a *Loop structure* can be executed multiple times. The number of times the loop executes depends on how it is configured: It could be set to loop only a set number of times, to loop forever, or to exit the loop when a particular sensor condition is reached, or similar.

NOTE A Loop checks only at the end of the sequence to see if it loops again (in terms of text programming, NXT-G uses a do while– or loop until–type construction, not a while loop), so any contained code is normally executed at least once.

A *Switch structure* is the NXT-G's version of an if-then. Based on some comparison that you specify in the configuration pane, one of the sequences contained within the Switch is selected to be executed. When you first drop a Switch, you'll see two possible branches within it: an upper sequence beam and a lower sequence beam. Only one of these branches will be executed, depending on how the start of the Switch block evaluates.

Simple, right? Well, there are a couple of very interesting complications. First, both of these composite structures follow the same rules for execution as a simple block. For instance, until they have completely finished execution (and that means executing all the blocks within them), they do not let execution continue on the sequence beam. They also follow all the rules for data dependency (discussed below).

tabbed view

There's a very nice (almost hidden) trick within the Switch structure: It can have more than one case! When first dropped, it has an either-or appearance: Either the top or the lower sequence executes. But there's an alternative way to view it. In the configuration pane, uncheck the box labled *Flat view*, and the Switch structure springs into an alternative tabbed form. Now instead of having both sequences displayed at the same time, they are arranged one behind the other, and you select which one is visible by using the small tabs along the top, as shown in Figure 2-6.

In this view, the Switch structure gains some new tricks. If you change the control pop-up to Value and the type pop-up to Number, along the right side of the configuration pane you will see a list of the possible options pop up. But now, the previously grayed-out + button is enabled; by clicking it, you can add new states to the Switch structure. So instead of executing just two possible sequences, it could execute a large number of different sequences, depending on what number is wired into it.

This actually works with the type set to Text as well, and there's even a "default" for this multicase Switch structure. Just select the state you want to make the default (execution will pass to this state if nothing matches the conditions you explicitly defined), and click the * button to make that case the default.

An even bigger advantage is that in tabbed view you can wire information into and out of the Switch structure. This enables you to use wiring more effectively to move information around in your programs. But before you start doing that sort of thing, you should probably learn a bit about wires.

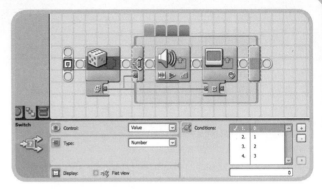

Figure 2-6: A multicase Switch structure, set to tabbed view and with extra cases added. Note that information can be wired across the Switch boundary in tabbed view.

NOTE Structures wait for every sequence inside them to complete before continuing or looping. This means that if more than one sequence is running in parallel within a loop, for instance, the loop will not cycle until all contained sequences have finished. If you're not aware of this, you might be surprised by how long it takes a simple loop to run, because in reality it is waiting for a slow sequence contained within it, including parallel spawned sequences. The exception is blocks set to not wait for completion (Sound, Motor set to "unlimited," etc.).

keeping things neat

One aspect of NXT-G that needs to be addressed eventually is how to organize and keep things neat and orderly. If you are not careful, complex programs can end up being a virtual haystack of wires, sequence beams, and other brightly colored distractions that can prevent you from clearly understanding your code.

wires

NXT-G uses *wires* as a way of transferring (or, more precisely, forwarding) information from one section of a program to another. As such they are very useful, convenient, and often overwhelming at first. There are some great topics covered in the built-in help, but one that is somewhat confusing is this: What's with all those wires? True, there's a lot in the help files on using wires, but often a new user goes through a progression of steps something like this: (1) programs are simple linear sequences of blocks and structures with no wires or very few; (2) user discovers wires in all their Technicolor spaghetti glory and starts trying to use them all over the place, perhaps also using branched sequence beams; (3) user becomes disillusioned with complex programs in NXT-G because they are a confusing mass of colored lines and often work in odd ways (or not at all).

There are at least two important things that can head off this unfortunate end to NXT-G programming: housecleaning (making your program understandable enough) and understanding (uncovering the rules that the wires operate by).

First, keep your programs well organized. Graphical languages can present a whole lot of information rapidly to the user, but if that information is presented in a confusing way, you'll confuse the reader that much faster. So keep things neat and well documented, even in a language that is essentially not text based.

To keep things neat (yes, I sound like your mother again . . . you'll get over it), pay attention to things like wiring. NXT-G actually tries to do an admirable job of this automatically. For instance, if you move the arrow tool over an output plug, it turns into a wiring tool, indicating that you could start a wire at this point. Click there and then move the wiring tool to an input plug ("drawing out" a new wire as you go); then click again to connect the wire.

Does it all look pretty, neat, and orderly? Maybe, but in a complex program, the answer is usually "probably not," as shown in Figure 2-7. You could make this neater by forcing NXT-G to organize the wiring for you; try to open and close some of the data hubs the wire passes by or under, or the data hub the wires starts from, and you can watch NXT-G attempt to "tidy up" the wiring—keeping wires from lying on top of one another or running under open data hubs that they don't actually connect to.

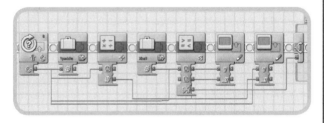

Figure 2-7: A fragment of NXT-G code with wires running all over the place. Notice how hard it is to determine where each wire is coming from or going to because of all the crossing wires and wires partially hidden behind data hubs.

manually organizing wires

Most of the time, the NXT-G environment's efforts are enough; it can actually untangle a Gordian knot of multicolored spaghetti reasonably well. But sometimes even that just doesn't work to your satisfaction. So what's a programmer to do? Do it yourself!

You can always manually define the path of a wire as you lay it down by clicking at intermediate points along the way. While this is a little more time consuming, it often produces a much more organized wiring system (with fewer bends and crossings) than if you allow NXT-G to do it on its own. Figure 2-8 is an example of this: It is functionally identical to the code in Figure 2-7, but is significantly easier to understand.

NOTE Save wire organization until one of the last steps in developing programs (or be prepared to rewire large stretches of code). If you open a data hub that might cross one of your wires, NXT-G will gleefully "improve" your hard work by shifting all the wiring into its own scheme. "Mother knows best" can sometimes get very frustrating, so save complex rewiring for a time when the code (or at least that section of it) is relatively finished.

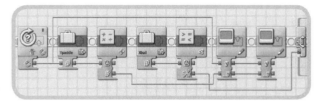

Figure 2-8: The code from Figure 2-7 "prettied up" to be comprehensible at a glance. By manually wiring out of the two-sided data plugs and carefully defining where the wires take right-angle bends, this improved diagram has only a single wire crossing (and one that has different colored wires to make it easy to distinguish).

MORE WIRING TIPS

There are lots of other wiring tricks you can use to clean things up. For instance, most of the time the initial wire has only one "floating" bend—sometimes to the upper right and sometimes to the lower left. By pressing the spacebar while you are still dragging the wire, you can toggle back and forth between these two options. You can also click and drag preexisting wires to try to straighten things up, or you can click the downstream plug to which it is attached to delete just one downstream section of a branched wire. There are other tricks as well—just play and discover.

comment tool

Another thing you can do to keep your code organized is to use the *Comment tool*, which is the little comic strip–like speech bubble thingy in the toolbar.

The fact that most folks forget about the Comment tool doesn't mean you shouldn't use it: Comment your code! Your code might have made perfect sense to you when you first created it, but invariably your memory will start to fade . . . a comment dropped in a critical place won't.

Comments are particularly helpful with long wires that stretch across the screen. For instance, instead of having to track back to the start of the wire to see what information it's carrying, you could just drop a label right below the wire, noting what information it's carrying. Or use comments with a series of Math blocks to remind you what they are doing. Sure, you can get all this information just by hovering over the blocks to display their configuration pane in the bottom left, but it's easier to understand complex expressions by reading the text placed above them, like the example shown in Figure 2-9. (This will come up again under topics like debugging, because comments are useful in many different situations.)

my blocks save time and simplify your programs

My Blocks are self-contained pieces of NXT-G code that you can define and "package" into a custom blue block that functions almost exactly like any other block in NXT-G. This way, a section of code you use again and again in your program can be reused, saving you from having to create it over and over, and making your program more readable . . . and *smaller*.

When an NXT-G program is compiled, the actual code used in a My Block is included only once. For each occurrence of a specific My Block in a program, the compiler just links to the single working copy of the code that is stored along with your program. This can result in a huge savings in memory on the actual NXT brick, not to mention much more readable, easier-to-edit code on the computer itself.

NOTE When looked at in terms of a text-based language, My Blocks are subroutines that can save memory and make your code much easier to understand. My Blocks are a win-win solution for a lot of programming issues in NXT-G.

Creating a My Block is remarkably simple in NXT-G. First, select a sequence of blocks on the programming sheet that you would like to "package" into a My Block. This package can contain variables, structures like Switches or Loops, and any basic blocks (it can even contain other My Blocks). Once you've selected the code you want, click the **=** symbol in the toolbar (or select **Make a New My Block** from the Edit menu). A dialog will pop up, showing you a small view of the code you've selected and asking you to name the new My Block (you can also include a description here if you wish). Clicking **Next** takes you to a second page of the dialog, where you get the chance to design a custom icon for the new My Block. Just drag and drop one or more icons from the palette below into the small display area near the top. Once placed, you can move these around by dragging them, or even resize them by using the small handles at the corners of the icons. When you are happy with the icon design, click **Finished**, and after a brief delay you will find that the string of selected blocks has been replaced with a single blue My Block.

This actually does far more than replace a bunch of code with a single simple block. The newly created My Block is now available just like any other block in NXT-G—from the pop-out palettes along the left side of the work area. Select the **Custom Palettes** tab and mouse over the top block; a pop-out palette containing any My Blocks you have created will spring out, and you can select any of these to drop into this (or any other) program wherever you might need them. In time you will build up a whole system of My Blocks, some very specific to certain programs and others of a more general nature that you can use again and again.

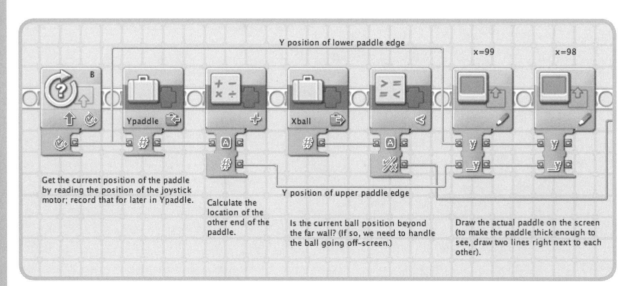

Figure 2-9: Using comments to explain what the program is doing

As a final point, if you had any data wires connecting your selected blocks to the rest of the program, these are not cut off, but instead are matched up with a series of plugs in the data hub that can extend from below the My Block. Use these plugs the same way you would on any other block—as a way to pass information into and out of it. If you want to see how these work within the My Block, just double-click a My Block in an existing programming sheet, and the hidden code will open up on a second programming sheet. From here you could edit it, add comments, or even rename the data plugs to something that makes more sense by using the Comment tool.

going with the flow: data dependency

Now back to those wires. The point of wires in NXT-G is to carry information from one block to others. Furthermore, different types of wires carry different types of information: Orange wires carry numbers, green wires carry logical values (*true/false*), and so on.

What a wire "carries" depends on where it is sourced. Wire initially has no valid value, not even a value of zero or an empty text string. Such an "invalid" wire will not deliver anything to its downstream plug until something is put into it, and once a value has been put into a wire, that wire carries that value to *all* its downstream ends, regardless of where they are in the program or what happens in the sequence between where the value was written out to the wire and where it is finally taken out of the wire.

For example, let's say you set a Variable block to a value of 3141, and then output that value into a wire. The wire itself now carries the value 3141 *even if the variable is subsequently changed* to something else (2718, for example). In essence, the value of the variable that the wire contains is now independent of the variable itself. This can be extremely useful (in a real sense, the wires are variables that you can set once but read multiple times), but it can be confusing if you think the wire connected to that Variable block will always stay in lock-step with the variable.

matching wires and plugs

If you've been playing with the NXT for any length of time, you probably already know that the wire type must match the type of plugs it connects to. For example, a plug that needs a number will not accept any wire other than a number wire. If the wire and plug don't match, the result is the dreaded "broken" gray wire (such as the ones shown in Figure 2-10), which is NXT-G's way of telling you that this particular configuration isn't going to work. (Technically, in

this case it signals a type mismatch: You are trying to connect an output plug of one type to an input plug of a different type.)

Another way of saying this is that wires can connect only plugs of identical types (Number, Text, or Logic), which may seem obvious. But why, then, does NXT-G signal broken wire in cases like these?

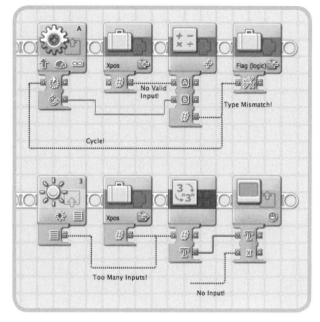

Figure 2-10: Two parallel sequences with a wealth of broken wire

Well, the simple (if somewhat unhelpful) answer is this: "Because it says so in the help files." Remember that I said reading those files would be helpful? Under *broken wires*, the help screen lists four possible causes:

1. Type mismatch: Input and output plug are of different types (for example, the wire near the end of the top sequence, commenting the Math block to the Flag variable).

2. Cycle: When you can follow a path of wires back to where you started (the long broken wire "backtracking" to the Motor block on the top sequence is an example).

3. Missing input: Those two-ended plugs, for example, need a proper incoming wire before the outgoing plug will work; no wiring out of a Variable block set to write, even though that output plug seems to just be sitting there to tempt you. (This is the problem with the wire leaving the Xpos Variable block, as well as the "hanging wire" near the bottom.)

4. Too many sources: Each wire can have only one source plug, although it can branch to connect to more than one destination plug (the wire trying to connect both the Light Sensor block and the Variable block into the Number to Text block has this problem).

Some of these rules (such as type mismatch) are probably self-explanatory: If a block needs a number and you try to wire in "four score and seven years ago," there will be some confusion. But the rest of the above cases hint at a deeper explanation, which is the dependency of blocks on the wires that feed them.

In fact, a deeper analysis of this situation shows that a block does *not* just blindly execute when its turn comes up on the sequence beam. Instead, it *executes only when all its inputs are valid*. Think of the sequence beam itself as one input (in fact, it's actually a special unique type of logical wire). For instance, if a Math block has two inputs, A and B (it actually has three: A, B, and the sequence beam it is hooked to), it will execute only when there is a valid value on both of those number wires (and the sequence beam has been "validated" by the completion of the block just upstream). If you understand this one principle about how NXT-G executes the graphical block code, a lot of previously odd or confusing aspects start to make sense.

With single-sequence programs, this really is a bit of a fine point. After all, most of the time all the wires carry information downstream from early blocks on the sequence beam to blocks farther along the sequence beam. But it certainly explains why a missing input (a wire with no source) results in a problem. If that wire connects to a block, when would that block execute? Since there is no way to feed a valid number into the "headless" wire, there is no way for this wire to ever carry a valid number, and any block that is *dependent* on it can never execute because it doesn't have all the information it needs. That one headless wire will hold up the whole flow of the program because it will wait for some information that can never come.

This also explains why a cycle (a wire sequence that can be followed back upon itself) isn't allowed. Where would the "first value" to be carried by this wire come from? It can't come from a block farther along the sequence beam; that block can't execute because the sequence hasn't yet reached it. At the same time, the sequence can never reach that block because the wire feeding into the earlier block can't have a valid value until the later block executes, and that block can only execute once the sequence progresses. But, of course, it can't progress because of the wire value . . . whew! Because such circular logic can't be logically expressed in NXT-G (with its data dependency), such wirings aren't allowed. (In fact, it's difficult to get a wire that is "broken" in this way—normally, NXT-G will not even let you draw it.)

a deeper look at dependency

The problem is that sometimes NXT-G will let you create a program that has all the wiring you think you want, but because of this dependency it has some very odd behaviors. For instance, say I want to control how long I stay in a loop based on a series of calculations such as comparing the current time to a randomly selected time, or even just a simple timed delay handled by a parallel sequence. Simple enough, right? The program might look something like Figure 2-11.

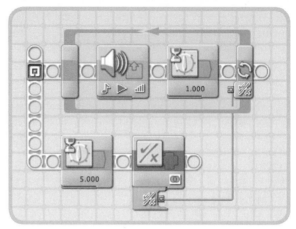

Figure 2-11: Control of a simple loop attempted with wires from a parallel sequence. The upper sequence should exit the loop only when the lower sequence, after a five-second delay, puts out a true on the data wire.

Now, if you try something like the program shown in Figure 2-11, you are likely to find that the upper loop does nothing until the lower loop finishes completely. As a result, instead of the upper loop clicking five times as it loops, it waits five seconds and then clicks once—as if the loop didn't loop at all, but just ran once right at the end of the five-second delay.

Why is this the case? Remember that the loop acts like a "normal" NXT-G block: It will not start executing until *all* wires running into it are valid. And until that lower sequence completes, the green logic wire running into the last logic plug in the loop will not be valid, so the loop "block" won't even start to execute. Remember when we discussed structures following the same data dependency rules as blocks? This is the consequence of it.

BOUNDARY PLUGS

If Switch and Loop structures correspond to blocks, then what corresponds to the data plugs where wires enter or leave? There actually are analogs of these, as you can see if you draw a wire across the boundary of a structure. As you do, you might notice that right where the wire crosses the boundary it looks a little odd. And in fact, for something like a Switch, if you click a section of a wire that crosses the boundary and press DELETE, only the wire up to the boundary will be deleted. This tells you that although the wire appears to be unbroken, it's actually connected to something. That something is called a *tunnel* in LabVIEW, and although tunnels are camouflaged in NXT-G, something similar is still present.

solutions using variables

So how can the example shown in Figure 2-11 be completed correctly? One way is to use variables. The Variable block can be used to keep track of things both inside and outside the loop, even as the loop executes. For example, you could have the loop check for the condition of an exit variable each time through, as shown in Figure 2-12. Now, every time the Variable block executes, it can put out the current value on the wire that feeds the loop condition, and the state of the variable itself can be updated by the second sequence running in parallel. In this case, the variable is initially set to *false*, and after a five-second delay the lower sequence sets it to *true*, kicking the upper sequence out of the loop the next time it reads the variable.

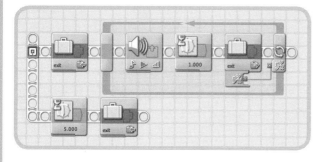

Figure 2-12: A solution to the problem shown in Figure 2-11, using a variable as a flag

using data dependency to your advantage

Data dependency might seem strange at first, but it can make some things much easier. For instance, it's hard (although not impossible) to do a calculation in NXT-G with the "wrong" or uninitialized values. Likewise, you can synchronize two parallel sequence beams using nothing more than a wire or two stretched between them, as shown in Figure 2-13.

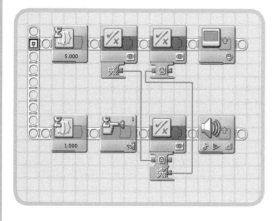

Figure 2-13: Two parallel sequences synchronized through data wires

In the figure, the upper beam sends a logical value to a block on the lower sequence. Only after the first wire is validated and the second sequence reaches that point can the lower Logic block execute, sending a logical signal back to the top beam, which has been "held up" by the lack of valid data on that wire. In this example, only the sequencing that you're using is important; what the actual Logic blocks are doing is completely irrelevant.

wiring into a wait: my blocks in action

Another odd problem or omission from the NXT-G block lexicon is that while there are Wait blocks, there is no way to wire a value into them. Try as you might, you will never be able to pull a set of data hubs from the bottom of a Wait block (even though it looks like you could).

Why not? The answer is probably to keep the interface clean and simple. For an awful lot of projects, you simply don't need the Wait duration under program control, and not having that option makes things a little easier to learn. But what if you're experienced and you want a Wait for *X* Seconds block, in which you can calculate and wire in a value of *X* seconds? Never fear—just make it yourself.

Here's where the expandability of NXT-G really becomes powerful. I can create a small piece of code that will function exactly as I want it to. For example, I could just create a loop that constantly checks the elapsed time since it was started and then compares that time to a value that the program can specify, as shown in Figure 2-14.

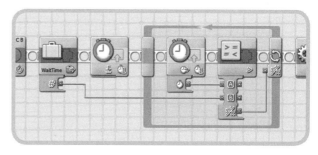

Figure 2-14: A custom "wait for it" type code

In this fragment of code, the time to wait can be specified by a variable (here called, imaginatively enough, *WaitTime*). Next, a timer is reset to zero, and the code starts to loop, repeatedly comparing the current elapsed time to the desired value in a Compare block. If the comparison is true (the elapsed time is more than the specified time), a *true* value will be sent out on the wire, tripping the code out of the loop (it is set to loop until *true*). Otherwise, the logic wire carries a *false*, and the loop executes again.

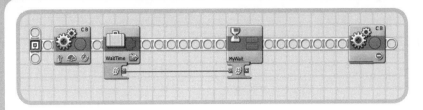

Figure 2-15: A My Block solution

a better way

This process works great, and we could use this piece of code anywhere we need it, re-creating it repeatedly, but there's a better way. Remember those My Blocks? By selecting the relevant code and clicking the = button in the toolbar, we can turn this section of code into a reusable subroutine: a My Block. After selecting the Timer block and the loop (no need to even select the stuff inside the loop—it will come along for the ride) and "ripping it" into a My Block, the code is immediately understandable and does just what we want it to: allows a value to be wired in to a subroutine that will do the wait operation for us, as shown in Figure 2-15.

NOTE *Ripping* **a My Block is a term that originated somewhere during the MINDSTORMS User Panel (MUP) and MINDSTORMS Development Panel (MDP) programs, when LEGO had a group of adult fans work with the beta versions of the software. Since no terminology had yet been laid down formally, a lot of informal terms were invented as needed by the group, including this one. Since programs seem to live on sheets of graph paper, making a My Block by highlighting a section of code and then "tearing it off" into its own sheet was how it was often talked about. While completely unofficial, it is rather descriptive.**

Notice that the wire that extended from the surrounding code into the My Block code has become a plug in the data hub of the My Block. As of now, this has the rather nondescriptive name *Value* (visible when the arrow cursor hovers over the plug), but we can change that to suit our needs.

Because a My Block is really just a stand-alone piece of NXT-G code, we can open its code for editing. Just double-click the My Block icon and it pops open into its own programming sheet, as shown in Figure 2-16.

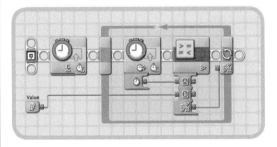

Figure 2-16: The My Block programming sheet for the MyWait block

Now, we can edit anything in the My Block that we want, including the label on the internal data plug (that funny-looking, stand-alone data plug near the start of the My Block code). You can have output plugs like this as well, if you created the My Block with wires extending out. We can use the Comment tool to make that label more descriptive—for example, *millisec*. (Since the NXT counts time in milliseconds in the Timer blocks, this is effectively the unit of time we need to wire in.) Once we've made that change and saved it, every instance of this block that we pull off of the My Block palette will have that new label.

And there's no need to stop there. You could edit a My Block to add blocks, or even add other My Blocks into the sequence: A My Block can include other My Blocks (or many of them, nested many layers deep).

cautions

There are a few cautions when using My Blocks, however. First, while it would be great to nest a My Block within *itself*, this is explicitly not allowed in NXT-G. (Sorry, there's no recursive programming with My Blocks; for the geeks among you, there is no stack analog that keeps track of local variables and return addresses.)

This also means that you can't efficiently run two copies of the same My Block concurrently: Instead of having a nervous breakdown, NXT will finish one instance of the My Block completely before executing the other one. Too, while any wire that enters or exits your My Block code obligingly appears as a plug in the completed My Block, there is, as of this writing, no way to add or remove such plugs *after* you've made the My Block. You can add new blocks, or even whole sequences to an existing My Block by editing its programming sheet, but not a new plug.

Still, if you really need a new plug, there are ways to create one. You could rip a new My Block directly out of the programming sheet of the old one, after adding another block with a wire so that the new My Block has an additional plug. Steve Hassenplug, one of the original four members of the MUP, was the first to point this out. It might seem obvious after you've tried it, but this is the sort of thing that's hard to do—thinking about things in new ways is tough.

NOTE Just as a sequence entering a structure does not continue beyond the structure until everything within it has finished executing, a program continues beyond a My Block only after every user-defined process within that My Block finishes. The data dependency rule that explains so much about the occasionally odd behavior of structures applies just as well to My Blocks. To put it another way, a My Block does not finish until all sequences spawned within it have concluded, the same way a Loop or Switch operates.

NXT-G problems and solutions

If the discussions in this chapter sometimes seem overwhelming, please do not be alarmed. The idea here is to try to give you a quick glimpse into some of the tricks and quirks of the NXT-G language that are not covered in the online help files. There's a lot covered in depth in these files, and I'm assuming that you've played around a little more than just getting your feet wet in the language. But if something herein doesn't make sense, let it go and perhaps come back to it later. It might be that you just haven't needed it. So, onto the sometimes puzzling world nearer to the heart of NXT-G.

While things like data dependency and the art of wiring are useful, there are still a few major problems (some subtle; some not so subtle) that will crop up from time to time. And while a single chapter can't cover all of the problems, there are a few that deserve immediate attention. Chief among them is wires. Again.

the dreaded "mystery" broken wire

Eventually, as you program larger and more complex pieces of NXT-G code, you will end up with a broken wire that warns when you try to compile and download your program. "No problem," you think, "I'll just go over my code one more time, find the wire, and fix it." But an hour later, having pored over the code, no broken wire can be found. There is no obvious gray wire anywhere on the screen, and you are about ready to chuck the whole computer out the window.

Don't (well, at least not yet); there are hints and a tool to help you. First, think of where broken wires may be hiding. One common place is under blocks, or under the borders of Switch or Loop structures. As they resize at will (when you add or delete blocks from inside them, for instance) or move around (as you add or delete blocks from in front of them), they can easily cover up a broken wire in a way that makes it very hard to find again. If you expand and contract or move chunks of your code, you might be able to find that sneaky broken wire.

But more often there's another problem: You can't shrink the structure enough to see what might be hiding behind it because, well, there's *stuff* in the structure. Even if it would let you shrink it (which it won't), doing so would likely make a bunch of other things hidden or broken. What's a programmer to do?

Thankfully, there's a helpful tool to fix this. Due to some incredible foresight on the part of the designers of NXT-G, not only is the language extendable with features such as My Blocks, but the programming environment is, as well.

remove bad wires

One of the tools that has been developed and can be installed into NXT-G has the name *Remove Bad Wires*. Originally authored by Jason King, this tool was the answer to many a programmer's prayer. Remove Bad Wires is an extension of the programming environment that just needs to be dropped into an internal folder of the application. Once there, it appears in the pull-down Tools menu as an item you can select. When you select it, it will simply go through and delete every broken wire in the program, saving you the trouble of finding and deleting them by hand.

NOTE This is not an officially sanctioned solution or addition to the MINDSTORMS software. It is a third party add-on; it has not been officially tested on lots of different systems, and if something goes wrong, there is no one to hold responsible for it (certainly not LEGO or NI, the folks who developed NXT-G). However, I have yet to see it fail, and it can make at least this troublesome experience go away in most cases.

Of course, before you can use this tool, you'll need to find and install it. You should find a copy floating about on the Internet somewhere, for instance at NXTasy (http://forums.nxtasy.org). To install it, do the following:

1. Create a folder or directory called *project* in the MINDSTORMS NXT engine folder or directory (if there's not already one there).

2. Drop the addition (called a VI) into it.

Now when you restart the environment, under the Tools menu you should see yet another nifty little tool, along with options such as Calibrate Sensors and Update NXT Firmware. Problem solved! If this doesn't work, you might want to read any help files or discussions that deal with such add-on tools. Books tend to take longer to go to print than the community does to develop and revise things.

running out of memory

When the NXT first came out, people were ecstatic over some of the new things it could do. Some of the limits imposed on the RCX (the RIS's version of the NXT brick) were simply blown completely away. For instance, instead of having just five program slots, the NXT allowed almost unlimited numbers of programs to be navigated with a full menu system! In place of a few canned sound effects and a frequency-selectable beep, the NXT could hold digitized sound fragments—finally, a LEGO robot that could speak! It also had a functional filesystem, stored images to display, and so on.

All these new features are wonderful, but they eat memory. And (strangely, in these days of cheap memory) the NXT seems to have only a paltry 60 or so kilobytes (KB) when you first start out. Why?!?

Well, first, memory is only cheap for your computer because it is designed to have expandable memory: The NXT is a small embedded device, and its memory isn't cheap. There is no room in its case to add memory expansion cards or a memory-management unit to access such cheap memory if it were installed.

NOTE Some people have suggested that since the NXT is a USB device, why not just change the style of the USB port to allow a plug-in thumb drive or similar device. The problem here is that there are really two different types of USB ports: *Master* (which supplies 5V of power in addition to controlling the communication) and *Slave* (the type implemented on the NXT, which does not supply 5V or have the on-chip code to coordinate a USB session on its own). While the externals of the NXT are literally so simple that a child can use it, the internals are a complex trade-off of space, complexity, and of course, cost. If the end result looks simple, that's a testament to all the people who carefully tried to design each aspect of the system, under some difficult constraints.

Sure, this capability could have been added, but remember that the NXT is designed to sell as a *toy*, not an infinitely expandable hobbyist robotics platform (that's just the way many of us use it). Also, while the NXT has a lot more than 60KB of memory installed, only a little of it is available to the user: Much of it is responsible for containing and servicing what is essentially a mini–operating system called the *firmware*. So while there is actually a good amount of memory in there, a lot of it is holding a program that is doing all the nice stuff that makes the NXT so easy to use, such as accessing menu systems, monitoring sensors, dealing with motors, and so on.

stretching the memory

While no one has (yet!) figured out how to add megabytes (MB) of fast memory to the NXT, you can stretch the existing memory.

To do so, first get rid of all the "extra" stuff. Deleting unused sound files, images, and programs, for instance, is one obvious way to free up some memory. When you connect to an NXT and look at how the memory is organized in the NXT window, you'll see the Delete All button, as shown in Figure 3-1. You can use this button to clear everything out (you'll have to load the stuff you want back on after this). But is that really everything?

Well, that's not everything. Note near the upper edge of the window a small checkbox labeled *Show System Files* (also shown in Figure 3-1).

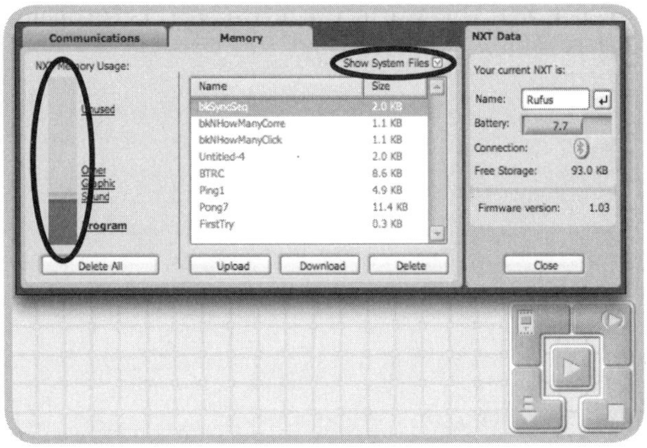

Figure 3-1: The NXT window, with the Show System Files checkbox and Memory map circled

If you check this box, you'll see that the bars along the right edge of the window suddenly jump up, suggesting that there are more programs, sound files, and so on in the NXT. You just revealed all the normally hidden system files—*all* of which are perfectly safe to delete!

While there are no critical files here, deleting these files might mean that certain options (such as on-brick programming or the Try Me programs) will be unavailable, but their disappearance won't crash the NXT. Best of all, deleting all those system files allows you to free up a little more than 120KB, almost doubling the available memory to hold your programs and sound files.

NECESSITY IS THE MOTHER OF INVENTION

While the NXT's memory limitations became apparent to a lot of folks at about the same time, FIRST LEGO League (FLL) teams ended up pioneering a lot of very inventive solutions very rapidly. As soon as they were available, many teams used Mini Blocks extensively (see below), but in some cases even that was not enough. A number of teams independently realized that they could save even more memory by running each mission with a single (very complex) My Block. They also realized that by putting all the missions in one superprogram, they could all share any common My Blocks, saving still more memory. They even ended up driving these superprograms with their own little menu systems, essentially changing in one season from a model of "different programs for different missions" to almost a state-machine approach that showed a lot of sophistication. All with the help of My Blocks, and a desire (and the creativity) to get around the memory limitations. Not bad for a bunch of 9-to-12-year-olds, huh?

If you want those Try Me programs and system sounds back, you can restore them as part of a firmware update simply by redownloading the firmware to your NXT.

NOTE A firmware update will restore these features, but it will erase any of the programs that you had on the brick as well!

Once you've cleared out all that memory, the best way to stretch it is to make sure not to fill it up too fast. One thing to watch out for is sound files. While these are a lot of fun, they also take up a lot of space. If you must have sound files, try to reuse them in multiple programs on the NXT brick (a single sound file can be used by more than one program, as a shared resource within the NXT filesystem). Another important thing to remember is that each instance of a My Block in a program does not duplicate a bunch of

code, but instead calls one reusable piece of code from multiple places. Again, this can save a significant amount of memory if used properly (for the geeks among you, My Blocks are compiled as subroutines, not in-line functions). Of course, yet another solution would be to simply use smaller versions of some of the standard blocks: Mini Blocks.

mini blocks

Mini Blocks, from NI, are a partial answer to this memory problem. They are a special set of upgraded blocks that you can add to the NXT-G language.

As it turns out, some blocks, such as the Move block, are doing complex things with a lot of options available to the user. That's great, but it means that before it will run, the code in the block needs to make sure everything is okay (wired-in values are in an acceptable range, and so on). That's a lot of checking, and some of it isn't needed in certain circumstances (if the program doesn't use those plugs, for example), so it can be streamlined out of the code.

NI also found ways to reduce the amount of code needed by other blocks. The result is a series of streamlined blocks that are smaller (requiring less memory), faster (less code to run through), and cheaper (hey, they're free; how much cheaper can you get?). Figure 3-2 shows a group of Mini Blocks mixed in with regular blocks.

Mini Blocks are an official release from NI and LEGO available from http://mindstorms.lego.com/support/updates. Browse at will (this is a good place to bookmark), but the file you want is called the Dynamic Block Update. Installing this patch enables you to import new blocks into the NXT-G environment, including the Mini Blocks and Legacy blocks (blocks that support the RCX sensors, with the addition of converter cables you need to order separately).

Once you have installed the Dynamic Block Update and restarted the software, you can download and install lots of additional blocks for NXT-G, including one developed by third-party distributors of various sensors as well as just generally helpful blocks other hackers have developed. (Do you want a compass or accelerometer for the NXT? It's available, and you can run it from NXT-G with a single block, just like the stock sensors.) To install a new block, open a programming sheet (you need to have some program open, even a blank one, for the Import tool to work).

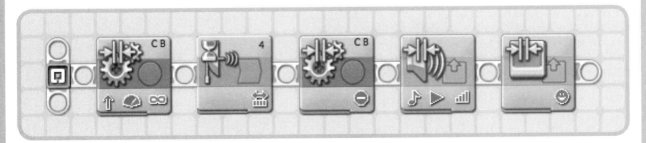

Figure 3-2: A short program with a mixture of Mini Blocks and stock blocks

Just select the **Import and Export Wizard**, and click the **Browse** button to direct the tool to the folder containing the block or blocks you wish to import. Once the wizard has listed all the blocks, select which blocks you wish to import and which palette you want them to appear in, and finally, click the **Import** button.

Once a new custom block has been installed, you can use it just like any other. Mini Blocks, for instance, work just as well as their "stock" brethren, but somewhat faster and significantly smaller (as much as 68 percent smaller than the first-generation blocks).

NOTE Many of these new blocks come from the LabVIEW Toolkit, a package available from NI. These are essentially most of the tools the development team at NI used to write the NXT-G environment in the first place, so in a way NI is offering its tools to the general LabVIEW programming population. This means all sorts of things are possible: blocks to handle text strings, do simple animations, perform complex numeric calculations, and so on. One of the "missing" features of NXT-G is the capability to handle arrays; but people have already developed new ways to use arrays in NXT-G by "rolling their own" blocks using this powerful tool.

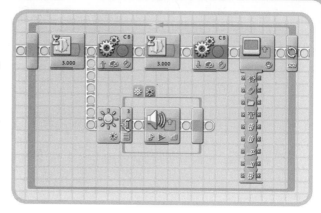

Figure 3-3: A crowbar opening up a loop. The Display block at the far right edge of the loop is not part of the final code, but just a temporary way to widen the loop by using a long extended data hub.

when to use a crowbar in programming

Advanced users of NXT-G encounter problems when attempting to construct a parallel sequence within a Loop or Switch structure. The idea is simple enough: Just drop an orphan block above or below the first sequence; then SHIFT-drag a branch of the sequence beam out to connect it.

The problem is space. Those darned structures are elastic and miserly: They spring down to the smallest size possible, yielding just enough screen space to show what is currently within them. So how do you get them to open up? You could politely ask them to "say ahhh," but it works much better if you use a crowbar.

the crowbar

In this context, a *crowbar* is just a kind of temporary tool you can use to open up elastic structures like this to give you room to start a parallel sequence. One simple way to do so is to drop a block with a really long data hub into the structure and then open the data hub by clicking on it. (The Display block works quite well for this purpose.) Now there's room to construct a parallel sequence below the first one, as you can see in Figure 3-3.

a bigger crowbar: the switch

If you need even more room than this, you need a bigger crowbar: the Switch. Since a Switch structure has a top and bottom part when first dropped, any structure that you drop a Switch into has to open up wide to accommodate it.

You can make that structure even wider (in a vertical sense) by putting something like a Display block on one of the two sequences in the flat view Switch, and then opening the data hubs on *that* block. The Display block expands the Switch (with both the top and bottom expanding to keep it symmetric), and the Switch forces open the outermost structure even more.

If you need still more room, you can even use another crowbar arrangement to open the first crowbar still more: By nesting crowbars, you can acquire just about any amount of screen space you need within a structure.

removing the crowbar

Once you've finished with your crowbars and constructed your parallel sequence beam, you can just select and delete the crowbar. (After all, this isn't code we want to compile; just a way to encourage the editor to make more room.)

The only problem is that the overzealous parent structure will now spring back to a small size, quite possibly hiding your newly constructed sequence(s). The sequences will compile fine, but to re-edit them you would have to go through the whole laborious crowbar process all over again.

One solution is to pin open the structure before you take out the crowbar. While the elastic boundaries of a structure don't always respect parallel sequence beams, they *do* respect comments. Therefore, to pin the structure open permanently, simply drop a comment near the bottom boundary. Now when you remove the crowbar, the structure boundary can snap back only to the pin you left in place, keeping it wedged wide open.

variables: the scoop on scoping

One last, somewhat advanced tidbit that can lead to confusion in NXT-G is variables. First it's important to realize that variables are not always required. Indeed, when you use wires properly to carry information from one place to another in a program, you'll find that you have less need to define "real" variables (the ones that need their own block) than you might think. Essentially, wires are temporary variables, and using them that way not only makes your code smaller and easier to read, but faster as well.

Still, those Variable blocks are in the palette for a reason, and there will be times when you need to use them. For example, when:

* Communicating information from outside a structure (Loop or Switch) to inside such a structure, to get around the data dependency issue.
* Casting a number (or other information) "far forward" along the line of the program. Many programs end up being very long and linear, and coaxing a wire to stretch all that way might simply be more trouble than it's worth.
* Transferring information deduced during one iteration of a loop back to be used in the next iteration.
* Communicating information into and out of executing a My Block.

local variables

The one cardinal rule for variables in NXT-G deals with a subject programmers call the *scope* of a variable. Traditionally, a variable declared in a subroutine *exists only in that subroutine*. It is, to use a common term, a *local variable*. Code outside of the subroutine can't look at or modify a subroutine's local variables, and code within a subroutine usually can't look at or modify variables declared in the main program or any of its other subroutines.

global variables

Global variables are ones that are defined outside of the main program and should be accessible from anywhere—whether from within the main program or from any of the program's subroutines. This means that a global variable changed in one subroutine can alter the way other subroutines that use that variable function, as well. For that reason, it usually makes good sense to use local variables because they allow subroutines to be portable. Subroutines using local variables are easily exchanged between different programs, because you know that the subroutine will never mess up another program's variables *even if they have the same name* (because of the local-versus-global dichotomy). And since My Blocks are an awful lot like subroutines, you might expect that to be the case

in NXT-G. More specifically, if you define a variable from within a My Block, you would expect that variable to be local. In other words, even if your My Block variable uses a common name (such as *X*), it shouldn't interfere with other My Blocks that might name one of their variables *X*. Right?

Wrong.

all variables are global

In NXT-G, it turns out that *every variable is global*. Any variable, defined by its name, can be accessed by any piece of code, either in the main program or in any My Block used within it. For example, if a program uses a variable named *Current angle*, any My Block that also has a Variable block labeled *Current angle* will reference that same variable.

One advantage is that you can use Variable blocks to pass information into a My Block that is already executing (something that is usually forbidden because of that data dependency issue) or even from one My Block to a different My Block that is running parallel to it.

the problem with global variables

The downside to having global variables is significant. For example, say you have a My Block that calculates a screen coordinate to plot something and uses a variable you defined and named *X* for some internal calculations. If the main program uses a variable called *X* as well, then every time it executes a copy of this My Block, the value of *X* could change in a way the main program wasn't expecting.

It's as if you and I were both sharing the same wallet to keep our money in. If you put $20 into the wallet, you expect there to still be $20 when you next open the wallet. You might be pleasantly surprised to discover $120 the next time (I cashed a paycheck), but greatly upset when you open it to find only a few cents in change (perhaps I bought some LEGO using the windfall I found in the wallet after *your* payday). See the problem? Variables, just like wallets or bank accounts, should be predictable.

solving the problem

What can you do about this? Well, first, use variables sparingly. If you don't have a lot of commonly named variables in your program and My Blocks, then this sort of conflict is less likely to happen. (As we discussed earlier, quite often a wire will work where you might normally be tempted to use a variable.)

If you must use variables in My Blocks, another way to solve this problem is to give them a name that is very unlikely to crop up somewhere else in the program. For example, if the My Block in question was named *ScreenCalc*, that variable previously referred to as *X* could instead use the label *ScreenCalc_X*. Sure, that's a mouthful, but by prefixing the variable name with the name of the My Block that uses it, you are much less likely to accidentally use it somewhere else (in another My Block, for example).

the remarkable untouchable variable

There's another cautionary point with respect to variables in programs and My Blocks: If you want full control of the Variable block, you must define each variable *by hand* for each programming sheet it appears in. But this doesn't mean a variable can't be used if it hasn't been defined in that sheet. Confused? Try this to reproduce the program shown in Figure 3-4:

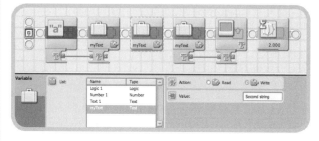

Figure 3-4: The original version of the program

1. Open a new program and define a variable named *myText* (with a data type of Text).

2. Drop a Text block with the A text string set to "First string" and then wire it into a Variable block to Write this to the variable myText.

3. Now drop a single Variable block to Write the string "Second string" to the myText variable.

4. Drop a Variable block set to Read that wires into a Display block set to draw the text to the LCD.

5. Follow this with a Wait block configured for two seconds (just to keep the display up long enough to read it).

When you run the resulting program, it should come as no surprise that the words *Second string* appear on the display.

But now try making the second Variable block in the program into the world's smallest My Block: Click just that Variable block and rip it into a My Block named *ChangeString* (shown in Figure 3-5).

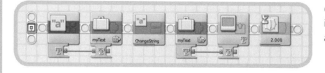

Figure 3-5: The same program with the middle Variable block ripped out and turned into a My Block

When you run this program, it produces exactly the same result; again, not really a surprise.

Now double-click the My Block icon to open up that code in its own programming sheet, and you'll see exactly what should be there: a single Variable block named *myText* (see Figure 3-6).

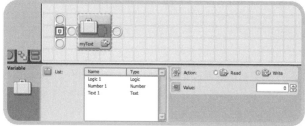

Figure 3-6: The My Block named ChangeString *opened up. Notice how even though the* Variable *block named* myText *is selected on the screen, the configuration panel seems to indicate that there is no available variable by that name, only the three default variables you always have.*

But if you click this Text block and look in the configuration pane, you'll notice something very odd: There is no variable named *myText* displayed in the variable list! NXT-G appears to have forgotten this variable, even though it is sitting right there in a named block on the screen, and even though the program works just fine! Furthermore, you'll find that you can't edit the value that you are writing to this block. (You can change it from Write to Read, though. What would you expect the program to do then?)

As bizarre as this behavior seems, the fix is quite simple. When the My Block programming sheet is open, select Define Variables from the Edit menu and define a "new" variable myText of a data type of Text. When you do, the Text block is again fully functional: You can edit it, changing its value to anything you like, and save it. After reinserting it in the program, it will behave just as you expect it to.

an invitation to continue

I'll finish the chapter by saying that I have not, by any means, covered all the quirks or tricks embedded within the NXT-G language. But this is a start.

Of course, we need robots to actually program the NXT-G, but even before you build it's good to know how to detect and fix errors in the code for those rare times when we humans just *might* make a mistake.

debugging—when the unexpected occurs

Programming has been described as an art as much as a science. We all have our own styles when it comes to programming NXT robots, and there really isn't any defined "correct" way to use the NXT-G graphical programming environment. One thing we will all have in common, however, are the inevitable pesky programs we create that just don't seem to behave or execute properly. When a bug is found during the execution of a program, it can sometimes be difficult to figure out exactly what is going on, let alone how to fix it. This chapter will give you some tips and tricks to make debugging your robots' programs a little easier and more productive.

This chapter will be broken into four sections, each one covering a method for making troubleshooting and debugging more effective. These methods are commenting, waiting with the Wait block, using the Sound block, and using visual clues with the Display block. When you finish this chapter, you should be better prepared to fix those troublesome programs that keep your robots from behaving properly and, ultimately, making you proud. Let's start with one of the easiest things to do when programming with NXT-G—adding comments.

care to comment?

Take a look at Figure 4-1. Do you have any idea what the NXT-G program is supposed to do?

You can probably tell a few things about this program just by looking at some of the blocks used. The Move block will obviously control one or more motors, but how many exactly? There's a Wait block configured to test the Light Sensor, but it's difficult to tell if it's configured to trigger for a well-lit room or a darker one. And what's the purpose of that Loop block?

As you can see, viewing an NXT-G program and determining its purpose can sometimes be very difficult—but it shouldn't have to be. Take a look at the same program in Figure 4-2.

Simply adding short text descriptions has made this program a lot easier to understand. Now we can see what a robot using this program will actually do. The program uses the Loop block to execute any blocks found inside it a total of three times. The Wait block will wait until the Light Sensor detects a low light reading (for example, if you cover the sensor with your hand), and then the Move block will spin motors B and C for a total of 180 degrees.

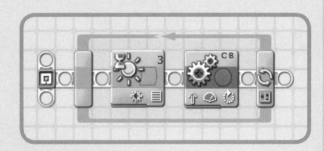

Figure 4-1: An unidentified NXT-G program

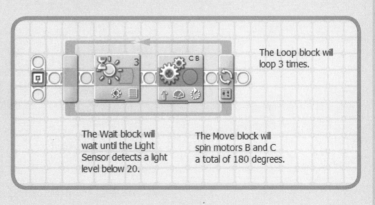

Figure 4-2: The same NXT-G program with comments

These short text descriptions are called *comments*, and they give a programmer the ability to describe various parts of a program using normal language. There are many advantages to using comments in a program.

First, comments are useful for reminding yourself about a program you might not have used or viewed recently. By reading the comments, you can quickly familiarize yourself with the purpose of the program as well as the individual components (blocks) and what they are supposed to do.

Second, comments are extremely helpful for others who wish to view your program and possibly learn from it. With detailed comments, a friend or even a total stranger should be able to take your program and be able to understand how it works.

Finally, comments can be useful for simply documenting your thought processes while programming. You can make notes in the program and also delete them if necessary, but they won't just get lost like some scribbling on a napkin or envelope might. Comments are a great way to keep track of various settings and tests you've run with the program, and they help you to avoid repeating mistakes.

Figure 4-3 shows the Comment tool on the NXT-G toolbar. If you click this button and then click anywhere in the program, you'll see a cursor and you can start typing. To start a new line, simply hit ENTER. When you're done typing, just click somewhere else on the screen and the comment will be added.

You can also drag and drop the comment to place it anywhere on the screen. Choose the Pointer tool, then click and hold on the comment and drag it around the screen. When you release the mouse button, the comment will change to editing mode and you can click anywhere in the comment to make changes or add new text. Clicking anywhere else on the screen will accept the comment and "lock" it in place.

The last thing you should know about the Comment tool is how useful it is for creating a summary of your program for anyone else viewing the program. Take a look at Figure 4-4, which shows the original program from Figure 4-1, now with detailed comments for the programming blocks as well as a program summary in the upper-left corner of the program.

If you include commenting in your programs, you will most likely find that debugging takes less time and effort. Develop a habit of adding comments for the NXT-G blocks you use and you'll quickly discover that it's not a lot of added work for the benefits you receive when trying to debug a troublesome program.

Next, I'll show you how to use the Wait block to help with troubleshooting and debugging programs.

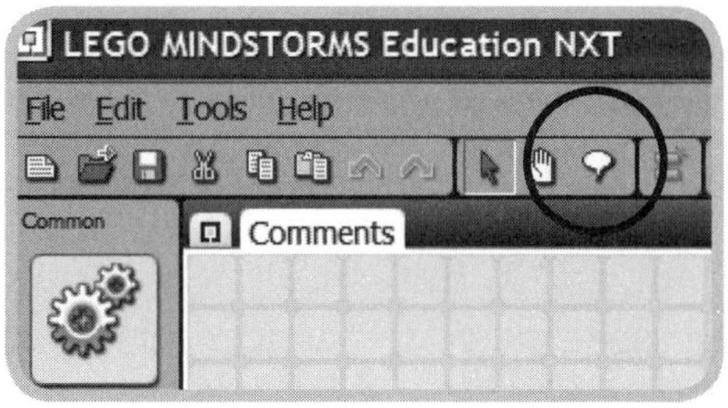

Figure 4-3: The Comment tool

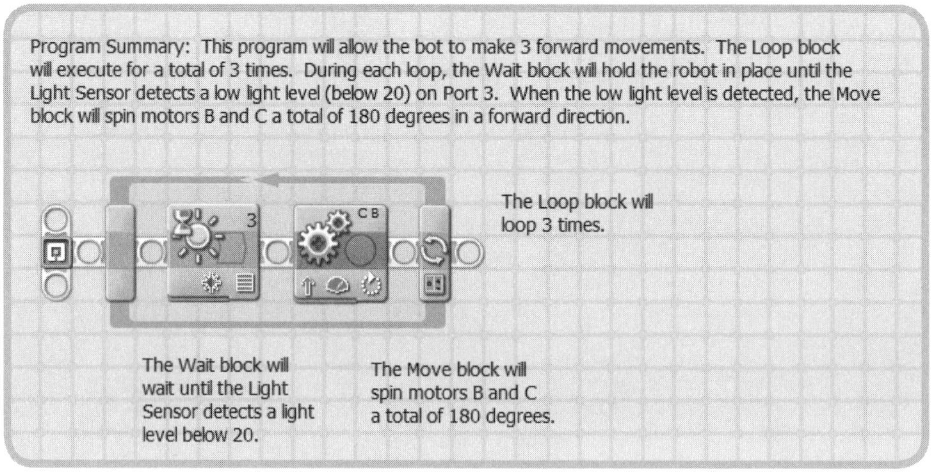

Program Summary: This program will allow the bot to make 3 forward movements. The Loop block will execute for a total of 3 times. During each loop, the Wait block will hold the robot in place until the Light Sensor detects a low light level (below 20) on Port 3. When the low light level is detected, the Move block will spin motors B and C a total of 180 degrees in a forward direction.

The Loop block will loop 3 times.

The Wait block will wait until the Light Sensor detects a light level below 20.

The Move block will spin motors B and C a total of 180 degrees.

Figure 4-4: A program with block comments and a program summary

waiting as a debugging tool

Figure 4-5 shows a very simple NXT-G program. Read the program summary and examine the block comments before reading further.

Here's a question for you: How can you verify that each of a robot's movements is correct before the robot continues on to the next movement?

If we run the program as shown in Figure 4-5, the robot will perform a total of seven movements and then return to its starting position. If one or more of the Move blocks isn't programmed properly, the robot will finish in a completely different location. What would be nice is if the robot could perform a movement and then let us verify that the movement is correct before it continues with the next movement. This would save battery power because we could stop the program if we encountered a problem, and the robot wouldn't have to run all seven movements (as well as the final Sound block). Well, there's good news, and then there's even better news!

The good news is we can use the Wait block to achieve this functionality. The better news is that there are different variations of the Wait block that give us even more debugging options. Let's go over a few here.

The first and easiest option is to include a simple Wait block that waits for a small amount of time before executing the next block. This amount of time should be small, but not too small—it should give you just enough time to watch the robot's movements and grab it and cancel the program if the movement is incorrect. This is demonstrated in Figure 4-6.

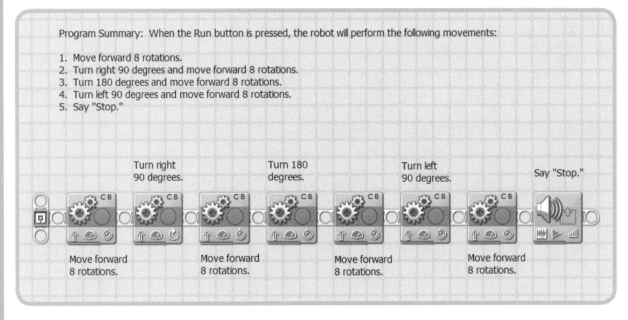

Figure 4-5: A simple program with comments

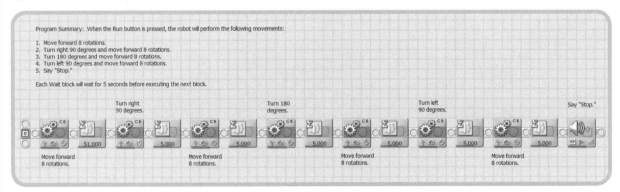

Figure 4-6: A program with simple Wait blocks

Each of the Wait blocks is configured to wait for five seconds. Feel free to bump the time up or down—the key is to give yourself enough time to verify that a program block does what you want it to do before the next block is executed. Later, when you are satisfied that the program works as planned, you can go back and delete the extra Wait blocks.

Another way to use the Wait block is to configure it with a user trigger. A *user trigger* is a command you give the Wait block telling it to wait until some sort of input from you, the user, is received. The two most popular methods for this type of Wait block involve using either the Touch Sensor or one of the buttons on the NXT brick.

Take a look at Figure 4-7. The program has been modified slightly by selecting *Sensor* from the Control drop-down menu on the Wait block's configuration panel. The default Sensor type is Touch Sensor, but you could select one of the other options, such as the Sound Sensor. (If you choose the Sound Sensor for the user trigger, you could use your voice instead of pressing the Touch Sensor button, but this might not work well in a loud room.)

You'll see the same program in Figure 4-8, but this one has the Wait blocks configured to sense the Left button being pressed. (Your other options for this are the Right button and the Enter button.)

When the Left button on the NXT brick is pressed, the Wait block will allow the program to continue on to the next block.

Using the Touch Sensor or one of the brick buttons as user trigger gives you a little more control than using a Wait block configured to wait five seconds. With these options, you can take your time verifying that your program and robot are performing properly; you might need to take measurements or verify rotation or direction values before letting the robot continue.

As you can see, using the Wait block can be a useful tool for debugging programs. This method can give you the time you need to watch and verify movements and other actions that your robot will perform. And, as mentioned earlier, when you're happy with the program, simply delete the Wait blocks and the program will run without waiting for a specified time or for user input.

Now that you understand how the Wait block can be useful for debugging, it won't be difficult for you to see how sound can be used as a tool for troubleshooting. In the next section, we'll cover some examples of using the Sound block when testing your programs.

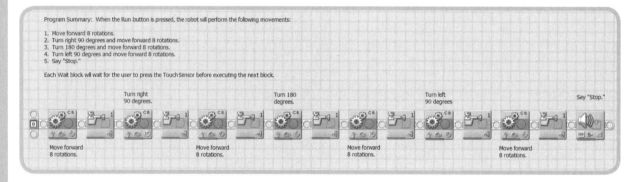

Figure 4-7: A program with Wait blocks configured to use the Touch Sensor

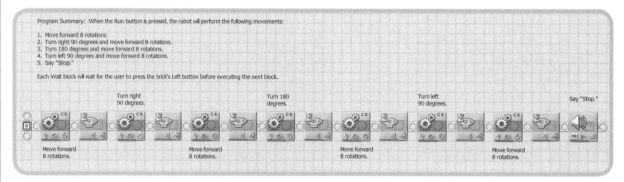

Figure 4-8: A program with the Wait blocks configured to use the Left button

listen as you troubleshoot

It's not a big jump to go from using Wait blocks to Sound blocks for troubleshooting. There are differences, however, that might not be immediately apparent.

First, a Wait block is going to wait at least one second (its default setting) before executing the next block. The Sound block, on the other hand, simply plays the sound you've selected without stopping the program.

Second, if the Wait block is configured to use the Touch Sensor or a brick button, your robot will sit there indefinitely until the user input is received or the batteries fail. The Sound block does not rely on any user input, so once the sound is played, the next block is executed.

You may be wondering how a Sound block can help you debug a program. For the answer, let's start with the small program shown in Figure 4-9.

If this program looks confusing, that's because it's supposed to. Many programs can become very confusing and tangled when loops are involved. That's why I've chosen to use Loop blocks to demonstrate the usefulness of the Sound block.

When running this program, it can become difficult to know exactly which block in the program your robot is currently executing. Is it running one of the innermost loops? Has it finished one of the outermost loops? As you can see, when your robot is running around, spinning motors A, B, and C, you'll have to pay careful attention if you want to know what's going on in your program. And that's where the use of the Sound block can help.

Take a look at Figure 4-10 and the modified program.

Can you see (or hear) what will happen when you run this program? Every time one of the outermost loops is completed, you'll hear your robot shout "Hooray!" When your robot completes one of the middle loops, it will either shout "Right!" or "Left!", depending on which of the innermost loops has completed. Unfortunately, this may not make sense until you actually use it, so I recommend creating a small robot (using all three motors with B and C used for motion) and uploading the program to watch it in action.

When running the program, if you've heard "Hooray!" twice and "Right!" once, you can probably determine that the innermost loops are running and motors B and C should be spinning your robot clockwise. You'll have to hear "Right!" two more times before you can expect to start hearing "Left!", because the middle loops are configured to execute three times.

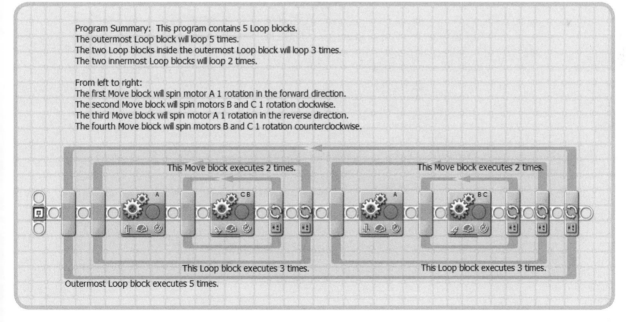

Program Summary: This program contains 5 Loop blocks.
The outermost Loop block will loop 5 times.
The two Loop blocks inside the outermost Loop block will loop 3 times.
The two innermost Loop blocks will loop 2 times.

From left to right:
The first Move block will spin motor A 1 rotation in the forward direction.
The second Move block will spin motors B and C 1 rotation clockwise.
The third Move block will spin motor A 1 rotation in the reverse direction.
The fourth Move block will spin motors B and C 1 rotation counterclockwise.

This Move block executes 2 times.   This Move block executes 2 times.

This Loop block executes 3 times.   This Loop block executes 3 times.

Outermost Loop block executes 5 times.

Figure 4-9: A program with a bunch of Loop blocks

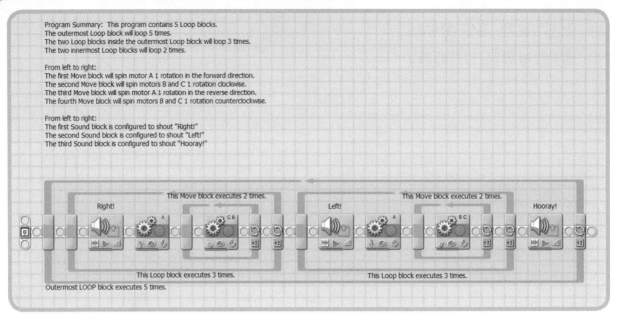

Program Summary: This program contains 5 Loop blocks.
The outermost Loop block will loop 5 times.
The two Loop blocks inside the outermost Loop block will loop 3 times.
The two innermost Loop blocks will loop 2 times.

From left to right:
The first Move block will spin motor A 1 rotation in the forward direction.
The second Move block will spin motors B and C 1 rotation clockwise.
The third Move block will spin motor A 1 rotation in the reverse direction.
The fourth Move block will spin motors B and C 1 rotation counterclockwise.

From left to right:
The first Sound block is configured to shout "Right!"
The second Sound block is configured to shout "Left!"
The third Sound block is configured to shout "Hooray!"

This Move block executes 2 times.   This Move block executes 2 times.

Right!   Left!   Hooray!

This Loop block executes 3 times.   This Loop block executes 3 times.

Outermost LOOP block executes 5 times.

Figure 4-10: A program with Sound and Loop blocks

Using Sound blocks this way will allow you to test your programs and provide sound clues to let you know which parts of the program are executing. If, for example, you never hear your robot shout "Left!", then you'll know you need to examine the part of the program that contains the second innermost loop and determine why it isn't executing.

Sound blocks can be extremely useful to you when you create an initial program. Be sure to choose different sounds for different parts of the program. As for placing a Sound block before or after specific blocks, this will come through experimentation. Some of us like to place Sound blocks before a specific action or NXT-G block as an alert of what is coming. Others of us like to place Sound blocks after specific NXT-G blocks have completed as an alert that something is done.

And remember that although Sound blocks don't take up a lot of memory on your NXT brick, it's still a good idea to delete the Sound blocks that are no longer useful to you to save space. You may find that leaving certain Sound blocks in the final program is useful, but just keep in mind that if a sound isn't absolutely necessary to the program's function, it can probably be deleted.

Now, if you've gathered that the Sound block can be a useful programming and debugging tool, you're going to enjoy this chapter's final debugging method. It is similar to the Sound block but can be used to provide much more information about your program and which block or blocks it is currently executing. What is it? It's the Display block, and it's covered next.

look before you leap

In the last section, I showed you how to use the Sound block to track the progress of your robot's program. By using sound, you can quickly determine which blocks in the program your robot is currently executing. Sound is useful for quick troubleshooting of programs, when you may only be concerned about whether your program is reaching or finishing a certain block. But sometimes you need more information, and that's where the Display block comes into play.

The Display block can put an image, some text, or a drawing (lines, points, and circles) on the LCD. Just like the Sound block, you can use information sent to the LCD as feedback. Instead of a sound, you can add Display blocks that will display something simple like a smiley face or other image. However, the real power of the Display block for troubleshooting comes from sending text to the LCD. This text can include things like numbers (from variables or sensors) or words (i.e., *Inner Loop Completed*) that will be useful to you when debugging.

For this last method, I'm going to use the simple program shown in Figure 4-11.

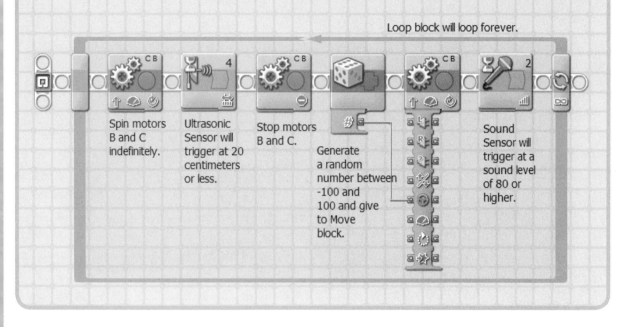

Program Summary: The first Move block will spin motors B and C forward for an unlimited duration until the Ultrasonic Sensor detects an obstacle (object, wall, etc.) within 20 centimeters. The second Move block will stop motors B and C and a Random block generates a number between -100 and 100. This random number will be provided to another Move block using a data wire and will control how far clockwise or counterclockwise the robot will turn. When the turn is completed, the Wait block will wait until it detects a sound level of 80 or higher before completing. The Loop block is set to loop forever, so the entire process will continue.

Loop block will loop forever.

Spin motors B and C indefinitely.

Ultrasonic Sensor will trigger at 20 centimeters or less.

Stop motors B and C.

Generate a random number between -100 and 100 and give to Move block.

Sound Sensor will trigger at a sound level of 80 or higher.

Figure 4-11: A program that allows your robot to wander around and avoid obstacles

The program is simple. Your robot will roll forward until the Ultrasonic Sensor detects an obstacle (i.e., a wall or other object), at which point it will stop. A Random block will then generate a number between –100 (negative) and 100 (positive) and send it via a data wire to another Move block that will either turn the robot clockwise or counterclockwise the specified turning distance. The robot will then wait for you to say something, snap your fingers, or make some other sufficiently loud sound before allowing the Loop block to start the entire program over again.

Notice that we are using the Sound Sensor as a means of troubleshooting. The robot will not move until the Loop block executes again, and the Loop block will only execute after the Sound Sensor has been triggered. We will eventually remove the Sound block once we've determined that the robot has been programmed

properly to randomly move around. But for now, we're using it as a means to test the randomness of the movement as well as to keep the robot out of trouble.

Now, here's a question for you: How could we track the Random block's generated values to make sure they are falling within the specified value range? It's easy—we'll use a Display block to send the value to the LCD for us to view.

Take a look at Figure 4-12 to see how to do this.

In Figure 4-12 we've added two blocks: a Number to Text block and a Display block. The random number generated is sent via a data wire into the Number to Text block, which converts the number to text. Then the Number to Text block sends the text to the Display block using another data wire, and the Display block displays the text on the LCD.

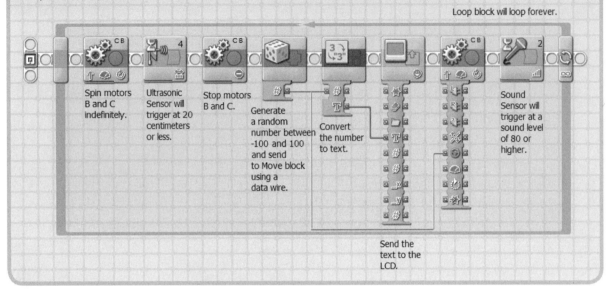

Program Summary: The first Move block will spin motors B and C forward for an unlimited duration until the Ultrasonic Sensor detects an obstacle (object, wall, etc.) within 20 centimeters. The second Move block will stop motors B and C and a Random block will generate a # from -100 to 100. This random number will be provided to another Move block using a data wire and will control how far clockwise or counterclockwise the robot will turn. This random number is also sent to a Number to Text block that converts the number to text and sends the text via a data wire to a Display block. When the turn is completed, the Wait block will wait until it detects a sound level of 80 or higher before completing. The Loop block is set to loop forever, so the entire process will continue.

Loop block will loop forever.

Spin motors B and C indefinitely.

Ultrasonic Sensor will trigger at 20 centimeters or less.

Stop motors B and C.

Generate a random number between -100 and 100 and send to Move block using a data wire.

Convert the number to text.

Send the text to the LCD.

Sound Sensor will trigger at a sound level of 80 or higher.

Figure 4-12: A program that shows us the random number generated

After you've run the program a few dozen times (or more) and are confident that the Random block is generating a number within the –100 to 100 range, you can remove the Number to Text, Display, and Wait blocks from the program. Now your robot will wander and explore, and you can rest assured knowing that the Ultrasonic Sensor is detecting objects properly and the robot is turning clockwise or counterclockwise a random distance before continuing on its way.

summary

When you create a program for your robot you're always hopeful that it will work on the first try, but this rarely happens. When your robot starts exhibiting strange behavior that wasn't part of the program, it's time to examine the program a little more carefully and figure out where the problem (or problems) are.

By using some of the most basic NXT-G blocks (Wait, Sound, Display) and adding detailed comments with the Comment tool, you can begin to track down those pesky bugs and help your robot perform its tasks properly.

making sense of sensors

While third-party manufacturers are hurrying to develop all kinds of sensors for use with the NXT, the basic Retail version comes with four external sensors and four internal sensors. In this chapter, we'll take a brief look at the sensors and, for your convenience, provide you with a reference chart identifying the various data plugs.

light sensor

Of all the sensors, the *Light Sensor* (see Figure 5-1) probably has the greatest variety of uses. A robot can use this sensor to detect light, follow a line, detect objects, and even scan images.

The Light Sensor can measure two different types of light: ambient light and reflected light.

* When measuring ambient light, the sensor simply reads the light intensity in its surroundings.
* To measure reflection, the sensor emits light from a red LED mounted above the input device and measures the intensity of the reflected light. If there's an object nearby, the light from the LED will be reflected and picked up by the sensor. The intensity of the reflected light is determined by how well the object reflects the light and how close it is to the sensor.

NOTE Although the Light Sensor has no built-in color detection, areas of different colors give rise to different numerical values of reflected light. After all, these are not absolute values; they are relative values that depend, for instance, on the level of ambient light. This makes color detection with the Light Sensor pretty unreliable and context dependent.

Figure 5-1: The Light Sensor

block types

Four different blocks can be used to make your robot react to readings from the Light Sensor: the Light Sensor block, the Wait block, the Loop block, and the Switch block.

The Light Sensor block (see Figure 5-2) is great for taking a reading from the Light Sensor and performing related tasks. The Light Sensor block's data hub outputs the Light Sensor's reading, either in the percentage of light or the raw numerical value of light intensity. It also outputs a logic signal, which is *true* or *false* depending on whether the reading from the sensor meets a certain condition that is set in the configuration panel.

As you can see, the configuration panels shown in all the blocks for the Light Sensor have an option at the bottom to generate light. Turning this option on makes the red LED emit light, which lets the sensor measure the light reflected by a nearby object.

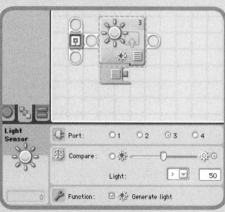

Figure 5-2: The Light Sensor block

The Wait block for the Light Sensor (see Figure 5-3) waits for the sensor's reading to meet a condition (which has been set in the configuration panel) before continuing with the rest of the program.

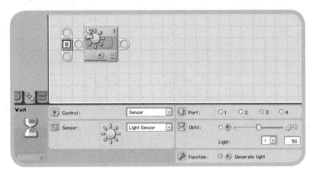

Figure 5-3: A Wait block for the Light Sensor

The Loop block for the Light Sensor (shown in Figure 5-4) is similar to the Wait block, in that it does something until a condition is met. Instead of just waiting like the Wait block, however, it continuously repeats whatever is inside its boundaries until the sensor's reading meets the condition. For example, if you put a Sound block inside a Loop block that has been configured to loop until the light reading is greater than 50, a sound will play until the light reading goes above 50.

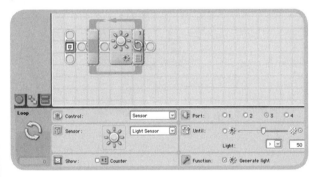

Figure 5-4: A Loop block for the Light Sensor

Finally, we have the Light Sensor Switch block (see Figure 5-5). This block takes the reading from a Light Sensor and compares it with a certain condition set in the configuration panel. If the reading meets the condition, the block or group of blocks in the top half of the Switch block will activate. Otherwise the block(s) in the bottom half will activate. In the example shown in Figure 5-5, the program will end if the reading from the Light Sensor is greater than 50. If the reading is less, motors B and C will be turned on.

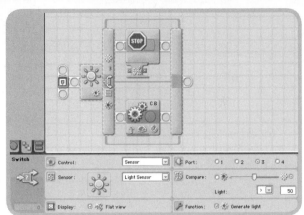

Figure 5-5: A Switch block for the Light Sensor

sound sensor

The *Sound Sensor* (see Figure 5-6) enables you to program your robot to respond to sound. It is often used as a trigger to start or stop an NXT action.

Figure 5-6: The Sound Sensor

If your robot seems to be turning itself off before you give it your sound cue, consider this:

* Noise from the robot itself or the environment can trigger the sensor. If the Sound Sensor is placed very close to the motors, or if the decibel level is set too low, the sensor might trigger at the wrong time. (The sensor does not discriminate between kinds of sounds; it simply measures the decibel level according to your settings.)
* If your sensor doesn't respond, make sure that the port you have chosen on the program block is the same as the port that your Sound Sensor is connected to on the robot itself.

block types

As with all the sensors, four different blocks can be used to detect readings from your sound sensor. The yellow Sound Sensor block (shown in Figure 5-7) enables you to connect to other program blocks with a data wire. This is useful when you want your robot to do something other than stop when it detects a sound cue. (We'll use data wires throughout the programs in this book.)

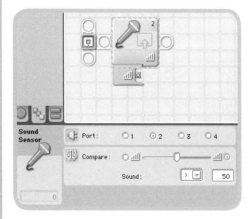

Figure 5-7: The Sound Sensor block

Wait blocks can be configured to halt the progression of your program until the designated sound is detected. For example, the program shown in Figure 5-8 will not progress until the sensor detects a sound louder than 50 decibels.

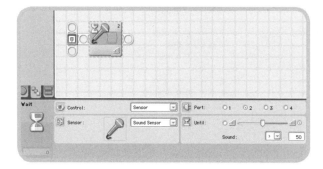

Figure 5-8: A Wait block for the Sound Sensor

BEWARE OF LOSING CUES IN A LOOP!

The Loop block (see Figure 5-9) is notorious for failing to stop action when the designated cue is given. For example, the Beach Buggy Chair in Chapter 10 will not respond to a sound cue when it is moving in a straight line, but it will respond when it is turning. If you're using sound loops in your programs, you might need to repeat the sound cue (as in, "Stop! Stop! Stop! Stop! Stop! . . .") until you determine at what point it will actually respond. When a loop is configured to depend on the value of a sensor, the sensor's value is actually read exactly at the end of a single run of the loop only; cues given while the blocks in the loop are processed are simply lost.

To overcome that limitation, you might be tempted to place some sensor blocks between the blocks of the loop, and wire their output to the loop's logic condition. Unfortunately, doing so will inflate your program unnecessarily with redundant code. A more elegant and reliable solution is to use a concurrent thread that contains a loop that (almost constantly) checks the sensor's value in very short intervals, breaks off once the desired event is received, and sets a particular logical variable. This variable can be wired to the logical condition of the (other) control loop, as shown in Figure 5-10.

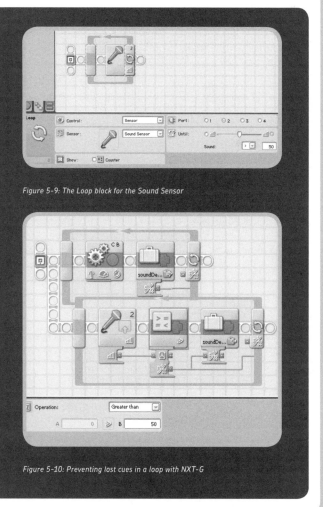

Figure 5-9: The Loop block for the Sound Sensor

Figure 5-10: Preventing lost cues in a loop with NXT-G

You can also program switches to respond to sound, as shown in Figure 5-11.

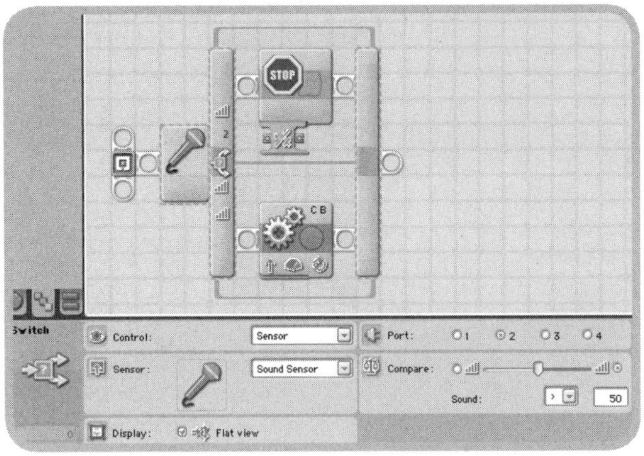

Figure 5-11: The Sound Sensor Switch block

This switch will cause a robot to stop completely if it detects a sound cue at this point in a program. In the absence of the sound cue, the motors will move forward.

The Sound Sensor detects sounds in two different modes:

* *Adjusted Decibels (dbA)*, which mimics the way the human ear measures ambient sound. In this mode the sensor ignores very low or very high frequencies.
* *Standard Decibels (db)*, which registers the whole frequency bandwidth equally.

The two modes can be configured using the *dBA* plug on the hub.

NOTE The decibel settings are set as percentages of a maximum value of 90 decibels, which is equivalent to standing next to a running lawnmower. Five percent or less is considered silence. A range from 5 to 10 percent would be equivalent to someone talking at a distance, while someone talking nearby or music set at a low volume would range from 10 to 30 percent.

what the sound sensor is not

The Sound Sensor's way of working is too coarse-grained to be used for voice recognition, so don't expect it to respond precisely to specific words or sounds. It responds only to decibel levels.

touch sensor

The Touch Sensor (see Figure 5-12) is rather straightforward.

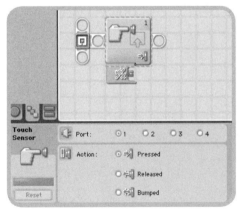

Figure 5-12: The Touch Sensor

As you can see in the configuration panel, the *Touch Sensor* detects when its orange button is pressed, released, or bumped, which makes this sensor the choice for detecting objects that the robot comes in physical contact with. (Note that the button is designed to accept an axle, which offers an expanded number of design options.)

block types

The Touch Sensor block (see Figure 5-13) responds when it receives a touch cue, but it does not halt the program until it detects that touch.

Figure 5-13: The Touch Sensor block

The Wait block configured to wait for the Touch Sensor (see Figure 5-14) causes your program to pause until it detects the designated action (Pressed, Released, or Bumped).

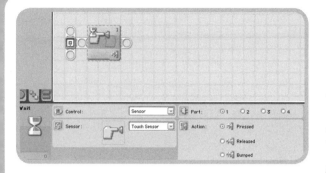

Figure 5-14: The Wait block for the Touch Sensor

In the Loop block (see Figure 5-15), the repetition is halted when the Touch Sensor is pressed.

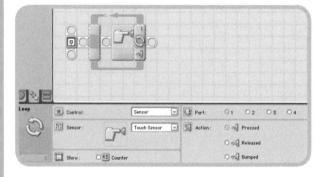

Figure 5-15: The Loop block for the Touch Sensor

The Switch block (see Figure 5-16) is configured to cause motors B and C to move when the button is pressed. When the button is released, everything stops.

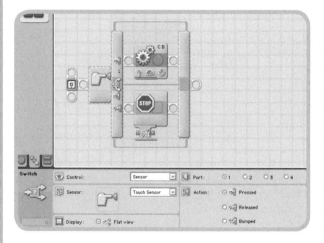

Figure 5-16: The Switch block for the Touch Sensor

ultrasonic sensor

The Ultrasonic Sensor (shown in Figure 5-17) is a newcomer to the LEGO MINDSTORMS universe (new with the NXT). Its prominent eye-like appearance already hints at its function: enabling a robot to see the world around it.

Figure 5-17: The Ultrasonic Sensor

But this is not your typical optical device; it doesn't actually process the photons in light. Instead, it works the way bats perceive their environment: by emitting high-frequency sound waves ("sonic signals" that are inaudible to the human ear) and processing their reflection from solid objects. The *Ultrasonic Sensor* can both recognize solid objects and compute their distance by evaluating the running time of a single signal.

Unlike the Touch Sensor, the Ultrasonic Sensor doesn't rely on physical contact with an object. Therefore, it can be used to detect and possibly avoid obstacles from a distance, before your robot runs into them. This ability can allow you to produce more lifelike robots that appear to see the world around them, rather than bump into it.

potential drawbacks

Depending on the way you mount this sensor to your robot and the kind of object it's sensing, the Ultrasonic Sensor's area of detection ranges up to 100 inches.

Yet, because of the nature of the signals it uses, the Ultrasonic Sensor has one great weakness: It cannot be used in an area in which another Ultrasonic Sensor is already at work because the signals sent out by the two sensors will interfere with each other and cause misreadings. In fact, because the signals sent out by the sensor are not uniquely labeled, one sensor would not even be able to distinguish reflections of its own signals from those of the other sensor.

Another potential pitfall when using this sensor stems from the unreliable reflection of signals by some types of surfaces. Round, very jagged, or soft surfaces are apt to disperse or swallow sonic waves.

accessing the ultrasonic sensor

To access the Ultrasonic Sensor in your NXT-G program, you will use the yellow Ultrasonic Sensor block shown in Figure 5-18.

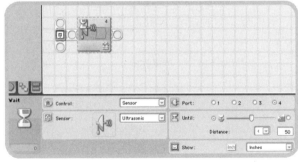

Figure 5-18: The Ultrasonic Sensor block

This block's configuration panel enables you to set the distance at which it triggers an event (measured in either inches or centimeters) and to define in which case the sensor signals: when it detects an object nearer than that distance or when it doesn't detect anything in this range.

wait block

To make your program pause until an object is detected inside of a given range (or to make it pause if no object is found there), you will use a Wait block (see Figure 5-19).

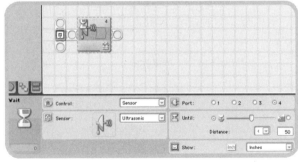

Figure 5-19: The Wait block for the Ultrasonic Sensor

In Figure 5-19 the Wait block is configured to be controlled by the Ultrasonic Sensor. This program will pause until an object is detected in a range of fewer than 50 inches.

You can use the Ultrasonic Sensor to evaluate break conditions of a loop, as shown in Figure 5-20.

In Figure 5-20 we configured the loop to break once an object is detected in a range fewer than 50 inches.

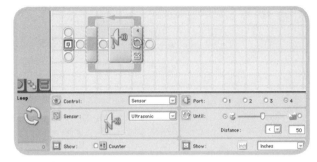

Figure 5-20: The Loop block configured with the Ultrasonic Sensor

switch conditions

You can also use the Ultrasonic Sensor for switch conditions (see Figure 5-21).

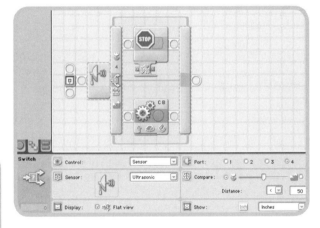

Figure 5-21: The Switch block for the Ultrasonic Sensor

In the example shown in Figure 5-21, the program will stop if an object is detected nearer than 50 inches. If not, motors B and C will move forward.

Always use the Ultrasonic Sensor with care because it can be unreliable in certain contexts. For example, when using it for break conditions in a Loop or a Wait block, your program might ultimately be trapped into a deadlock if the sensor does not detect the object as desired.

Still, for most scenarios, the Ultrasonic Sensor offers a very powerful way to control a robot and it is certainly one of the most valuable recent additions to the LEGO MINDSTORMS world.

built-in sensors

Your MINDSTORMS NXT kit contains four internal sensors that you will see listed whenever Sensor pull-down menus appear. For example, Figure 5-22 displays the internal sensors as shown in the palette.

Figure 5-22: Internal sensors shown in the palette

Figure 5-23 shows them in the Wait block configuration panel.

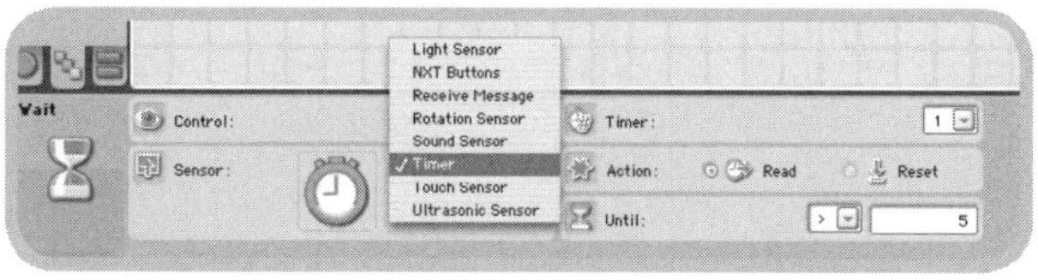

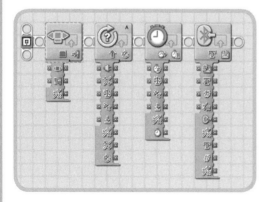

Figure 5-23: Internal sensors shown in the Wait block configuration panel

Figure 5-24 shows them in the Loop block configuration panel.

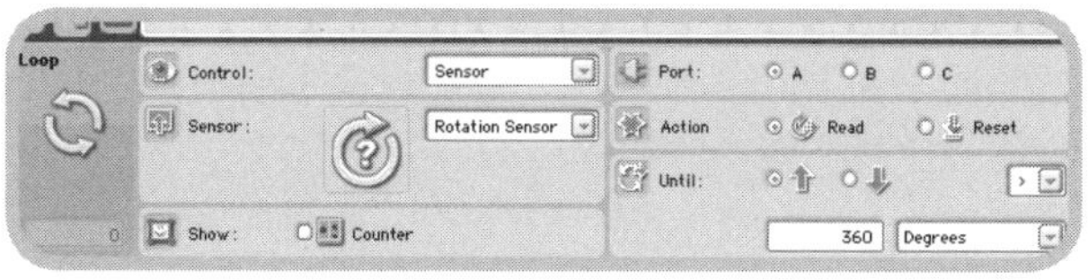

Figure 5-24: Internal sensors shown in the Loop block configuration panel

The four internal sensors are as follows: NXT Buttons Sensor, Rotation Sensor, Timer Sensor, and Received Message Sensor (see Figure 5-25).

Figure 5-25: Built-in sensors

The *NXT Buttons Sensor* enables the buttons on the NXT brick to be used in your programming. The Left, Right, and Enter buttons can be programmed to respond to one of three different actions: when the button is pressed (pushed in), when it is released (after pressing), or when it is bumped (both pressed and released), as shown in Figure 5-26.

Figure 5-26: The NXT Buttons Sensor

The *Rotation Sensor* enables you to sense the rotations of your motors by using degrees or the number of rotations, as shown in Figure 5-27. This sensor is used extensively and exclusively with motors.

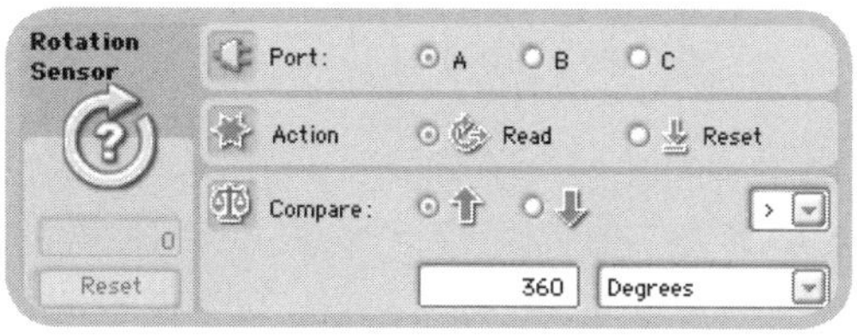

Figure 5-27: The Rotation Sensor

The *Timer Sensor* enables you to use the NXT internal clock. It can read the internal clock to determine how much time has passed and can be reset as a part of your program, as shown in Figure 5-28.

Figure 5-28: The Timer Sensor

The *Receive Message Sensor* (shown in Figure 5-29) is primarily used for wireless communication to allow your robot to respond

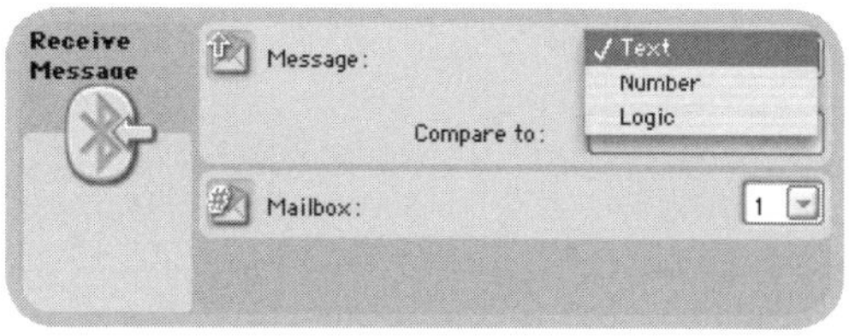

Figure 5-29: The Receive Message Sensor

to Bluetooth messages sent by a remote device (like a computer, a cell phone, or another NXT brick).

It enables you to compare the received messages by data type (Text, Number, or Logic) as well as values sent to a designated mailbox. These features allow it to differentiate between senders and types of messages.

data plug reference chart

Table 5-1 shows a quick-reference list of all the different data plug symbols used in the sensor data hubs. If you don't use data wires often, this list will provide a convenient reminder of all your options.

table 5-1: data plug symbols used in sensor data hubs

data plug symbols	data categories	data type/limits
	Action	Number (0 = pressed, 1 = released, 3 = select)
	Button	Number (1 = right, 2 = left, 3 = select)
	Degrees	Number (1–2,147,483,647)
	Direction	Logic (true = backward, false = forward)
	Distance	Number (0–255 cm, 0–100 in)
	Generate light	Logic (true/false, Light Sensor)
	Greater than/Less than	Logic (true/false)
	Intensity	Number (value read from Light Sensor)
	Logic in	Logic (true/false)
	Logic out	Logic (true/false)

table 5-1: data plug symbols used in sensor data hubs (continued)

data plug symbols	data categories	data type/limits
	Mailbox	Number (1–10)
	Message received	Logic (true/false)
	Number in	Number (–2,147,483,648–2,147,483,647)
	Number out	Number (–2,147,483,648–2,147,483,647)
	Port	Number of input port (1,2,3, or 4)
	Raw value	Number (0–1,024)
	Reset	Logic (true = reset, false = read)
	Timer value	Number (1 = timer 1, 2 = timer 2, 3 = timer 3)
	Trigger point	Number (0–2,147,483,647)
	Yes/No	Logic (true/false)
	Text out	58-character maximum
	Text in	58-character maximum

6

design

Designing a robot from scratch may involve more art than science. It is no surprise that the most famous inventor, Leonardo da Vinci, created both artwork and mechanical designs. Although he is well known and admired for his paintings, it is his notes about, theories on, and sketches of machines that have inspired many modern-day technological designs.

He believed that the key to creating new inventions was to separate existing machine designs into smaller, reusable, functional parts, which could then be examined in detail. By gaining a greater understanding of how each individual part functioned, he was able to modify the parts and easily envision how combining them in different ways could result in new inventions. He was the first to realize that the creation of new machine designs involved learning as much as possible about individual parts and then imagining how part functions could be used together in real-world tasks.

Although the LEGO NXT did not exist back in da Vinci's times, it still fully qualifies as a working machine under his definition. He recommended designing with small, simple, and reusable parts, just like the ones found in the LEGO NXT kit—so why not apply some of his guidelines to your own robot design process? That is just what we will do in this chapter. The following sections provide guidelines, sample worksheets, and blank worksheets to assist you in taking inventory of the hardware, software, and LEGO building parts available for your designs. The worksheets also allow you to more easily document how resources contained in your inventory can function together as robotic tasks. By organizing known data, as well as your findings and observations, you will gain knowledge about how each robot part works and improve your chances of inventing clever designs.

taking inventory of available resources

Although it is possible to build robots that are able to perform advanced behaviors and accomplish complex tasks, your designs are limited to the hardware, software, and building parts available to you. For example, a design concept may require a robot to navigate its environment using some type of vision system. An ideal design may implement expensive components such as stereo cameras and laser range finders, neither of which is available as a LEGO NXT part. Most often, hardware unavailability or high costs prevent designers from implementing a desired robotic function to full expectations or to the potential of existing technology. In most cases, a scaled-down design is used as an alternative. For example, you may consider using two LEGO NXT Ultrasonic Sensors to avoid obstacles, as opposed to using advanced vision. Of course, using simpler and cheaper solutions generally limits how well a robot can make decisions. A robot using advanced vision could easily navigate around an office building without getting stuck. However, this would not be possible using only two Ultrasonic Sensors. It is important to learn about how operating ranges and limitations for each resource in your inventory may potentially affect the scope of a robot's design. A robot designer needs to reach a good understanding of how each individual part works, how it communicates with other parts, and the role it can play in a larger design.

The following sections offer worksheet examples to help you track your design resources, as well as guidelines for designing and tips and techniques to help with the design process. It's a good idea to keep your worksheets updated with resources you obtain or test over time. The information on your worksheets can be reused in different ways for future projects.

taking inventory of smart parts

The first step to organizing your resources is to take an actual count or inventory of which hardware parts you have available for your designs. In order to clarify the process of mixing and matching parts in different combinations, we will categorize hardware resources into smart parts and building parts.

A resource qualifies as a *smart part* if it has some built-in electronic function or can communicate with other parts. For example, a LEGO piece with a built-in lamp qualifies as smart, but a simple LEGO building piece would not. Smart parts include electronics, robot peripherals, sensors, motors, joysticks, and even laptops. These parts do not necessarily need to be manufactured by LEGO, so mindsensors or HiTechnic sensors would also qualify.

It may also be helpful to include functional parts that may be physically attached to others. For example, the NXT brick has a built-in LCD, four buttons, as well as a speaker, which could each be listed separately as smart parts.

For now, you can list the most interesting parts you already own, such as sensors, actuators, or motors. If you have third-party sensors such as a compass or accelerometer, add those as well.

Just list a few in the *Parts* column of a blank Smart Parts Worksheet, and we will add more detail for each part in later sections.

See the sample Smart Parts Worksheet for a few example parts.

listing smart part functions

Once you have inventoried your most interesting smart parts, you can make a list of useful ways each part can perform functions. A *function* can be any operation, calculation, or behavior that can be used in combination with other functions to complete a robotic task.

smart parts worksheet (sample)

part	qty.	native part functions	software-enhanced part functions	known limitations
Ultrasonic Sensor	2	Measure distance Detect motion Detect an object's proximity Get analog input from a user	Follow a wall Detect stairs Detect motion over time Follow an object	10 × 2 foot range Cross-talk with other Ultrasonic Sensors
Touch Sensor	1	Detect physical contact with an object Detect collisions Get user input as a button Trip a switch	Determine how often a switch is tripped Determine at what time a switch is tripped	Requires some pressure to activate
Light Sensor	1	Detect brightness Detect motion	Detect colors Detect brightness change over time	Short range Susceptible to environment lighting
Motor	3	Position an axle, arm, or wheel Rotate a wheel Rotate a gear Get analog input from a user as a turn knob	Precisely position a servo Keep speed constant Move at a set speed of meters per second	Limited rotational torque Limited speed Bulky
Internal Rotation Sensor (built into motors)	3	Track a wheel or axle's rotation count Track a wheel or axle's rotational direction Detect the rotational movement of a wheel or axle	Control speed using proportional integral derivative (PID) C code Balance robot Measure distance traveled Control traction Navigate using dead reckoning	Ticks are missed if wheels turn too quickly
Compass	1	Detect compass heading	Orient a robot	Metal and magnetic fields interfere with digital compass Requires calibration to longitude/latitude
Sound Sensor	1	Detect the presence of sound in a room	Trigger a voice command	Susceptible to background noise

You may need to experiment with parts individually to learn how they work.

First list which native functions each part can perform out of the box and as specified by its manufacturer. *Native functions* are those that an individual part can independently perform without relying on other parts. Once a part's native functions are known, we can find interesting ways in which they can be enhanced with software, and we should list those as well.

Example 1

Native functions for an Ultrasonic Sensor might possibly include the ability to measure distance, detect motion, detect object proximity, or act as an analog input device.

Software-enhanced functions for the Ultrasonic Sensor might include logging detected motion over time, taking distance measurements in feet or meters, or detecting proximity within specific ranges.

Example 2

Native functions for a Touch Sensor might possibly include the ability to detect a collision, detect contact with an object, or act as an input button or trip switch.

Software-enhanced functions for the Touch Sensor might include the inputting and recording of a musical beat, alarm snooze button, or delayed trip switch.

Example 3

Native functions for a Light Sensor could possibly include the ability to detect brightness, detect day and night, detect lights turning on and off, or detect motion within a close range.

Software-enhanced functions for the Light Sensor could possibly include detecting colors, following a line, reading a bar code, or scanning an image.

Go ahead and write down some function ideas for your inventory using the Smart Parts Worksheet. There is a column provided for both native and software-enhanced functions.

listing smart part limitations

There may be potential limitations to prevent you from building a robot design. Each part in your inventory has specifications and operating thresholds that should be understood in order to define a design scope. You may refer to the manufacturer's documentation for each part and note relevant limitations for each function on the worksheet. Knowing the functional limits ahead of time might prevent unexpected or disappointing results. For example, an Ultrasonic Sensor can only measure up to a certain distance.

Other undocumented or less obvious restrictions such as motor torque, memory size, graphic resolution, and software bugs will affect your designs, and may only be discovered over the course of the design process.

Some LEGO NXT part limitations include a maximum payload capacity, maximum speed, maximum number of sensor ports, and little memory space for sound files. Now is a good time to note some limitations for functions you listed in the Smart Parts Worksheet.

combining smart parts into groups

Now that we have a list of parts to pool from, we can find different ways they can be grouped together to obtain more complex functions.

For example, the Ultrasonic and Touch Sensors can be combined to obtain a collision avoidance and detection system.

In order to facilitate the design process, groups of parts are listed on a new Smart Part Groups worksheet. If you have some initial ideas about which parts in your inventory could be grouped together, list them now in the *Part Groups* column.

listing smart part group functions

We have listed each part's native functions in the previous worksheet, so we don't need to repeat this process for groups of parts. However, we can find interesting ways in which the groupings can be enhanced with software and add them to the worksheet.

For example, using two Ultrasonic Sensors together with additional software could allow the robot to determine if an object is to its right or left and navigate an appropriate path.

Initially, it may not be obvious which parts can be combined together, but this will become clearer as you learn about how each one works.

See the sample Smart Part Groups Worksheet for function examples.

smart part groups worksheet (sample)

part group	qty.	software-enhanced functions	known limitations
Ultrasonic Sensor Touch Sensor	1	Detect and avoid collisions	Robots can be a maximum of 9 inches tall and 9 inches wide
2 Ultrasonic Sensors 2 Touch Sensors	1	Detect and avoid collisions Determine if obstacle is to the left or right of robot	Requires additional sensors

listing smart part group limitations

The limitations for each part in a group have already been listed in the previous worksheet, so we won't repeat the process here.

It is important to note that grouping parts together may introduce new limitations when they function simultaneously. These limitations may not be discovered initially, so you can come back at a later time and add notes to the worksheet in the *Known Limitations* column.

taking inventory of LEGO building parts

In the first few sections, we worked with hardware such as sensors, which were categorized as being smart. The second category of hardware resource contains LEGO parts used for building the chassis or mechanical structure of the robot itself. For example, a wheel, gear, axle, and LEGO beams or blocks all qualify as *building parts*.

listing individual building parts

Now is a good time to go through your LEGO kits and set aside any building part that you think may be used to perform a mechanical function or play a role in a robotic task. Parts like bricks and beams are not very unique or mechanically functional, so we won't bother listing all of those. When you identify some, you can list them using a new Building Parts and Assemblies worksheet.

listing assemblies

An *assembly* is a group of LEGO parts that performs a mechanical function as a whole.

For example, the Alpha Rex walking robot includes both leg and arm assemblies, either of which could be used separately in other designs. The legs could potentially be used to design a walking robot with a different upper body, and the arms could be mounted to a six-legged chassis.

Another example is the stinger found in LEGO's Spike design. This assembly can be built with the standard parts that come with the MINDSTORMS kit, so it is surely a good candidate to be reused for other designs.

The NXT-G tutorials and build instructions are good sources for ideas on the assemblies you initially have available.

Some examples have been added to the sample Building Parts and Assemblies worksheet.

NOTE In some cases, a part may qualify as both a building and smart part. An example of this would be a differential gear box with built-in shaft Rotation Sensor (although there are currently no parts of this type in the LEGO NXT kits). It is okay to list a part in more than one category.

listing LEGO part and assembly functions

Once your most interesting building parts have been inventoried, you can make a list of useful ways in which each part's mechanical functions can contribute to a robotic task.

For example, a gripper assembly can be used to pick up a ball, shake a hand, hold an opponent, or draw with a marker.

If you can think of different ways the stinger or other building parts could be used, list them in the *Mechanical Functions* column on the worksheet.

building parts and assemblies worksheet (sample)

building part or assembly	mechanical functions	known limitations
Claws	Pick up ball Grab opponents Hold an object in place	Limited torque
Gripper	Manipulate object Turn a knob Pick up an object	Limited torque
Axle differential	Control traction Use a single motor to drive two wheels	
Stinger	Attack opponents	
Legs	Walk	Limited movement Limited turning capability Cannot walk over obstacles
Arms	Position sensors	Limited freedom of movement

listing LEGO part and assembly limitations

Each building part in your inventory has mechanical or structural operating thresholds that should be considered. Some parts may not be strong or large enough to complete a required task. For example, a LEGO robot could easily pick up a ping-pong ball, but not a basketball. Tasks such as climbing stairs and traversing outdoor terrains require heavier-grade materials than what LEGO parts can provide.

Some building part limitations are undocumented and more difficult to quantify. Most structural limitations are observed during the testing phase of a design.

taking inventory of chassis types

The *chassis* is the basic frame or structure of a robot to which all other sensors or components are attached. The chassis type used will determine the robot's payload, stability, and ability to operate within its expected environment. For example, a chassis for a two-legged walking robot may not have as much payload capacity, stability, or the low center of gravity of a four-wheeled design.

Different chassis types can perform specialized functions. It is a good idea to record the different chassis types you are familiar with, along with the kinds of functions you think they may be used for. Over time, you will come across and build different chassis types that can be reused for different designs.

It may not always be possible to detail building instructions for each chassis, but you may want to, at minimum, note the key parts used to assemble it, along with the functions it can perform.

Also, a chassis can use many parts, so you may only have one built at a time. If you can, you may want to make some notes while it is intact.

listing chassis functions

Different chassis types can perform specialized functions. Some designs may be more functional than others when it comes to operations such as turning in a tight radius, carrying payloads, or climbing over obstacles.

Not every chassis is a vehicle or has wheels. For example, a robot arm base cannot drive around a floor, but it still qualifies as a chassis.

You can use the Chassis Types worksheet to list some you are already familiar with. A few are fairly obvious, but as with building parts, it is up to you to think of ways in which a chassis's functions can be applied to robotic tasks.

chassis types worksheet (sample)

chassis type	key resources	functions	known limitations
Two-wheeled base	2 motors 1 accelerometer Rotation Sensor	Turn in tight spaces Turn around in a corner Keep payload and sensors level	Needs complex PID-balancing software and custom not possible in NXT-G Needs balanced center of gravity Some designs require accelerometer sensor
Three-wheeled base	2 motors	Easily turn in tight corners Drive up ramps and over small bumps while both drive wheels remain in contact with floor	Not as much traction as tank treads Limited stability requiring low center of balance
Four-wheeled base	2 motors Steering assembly	Carry a heavier payload and keep it stable Provide a low center of balance for a payload	Turning and dead-reckoning software is more complex Requires steering mechanism Carry heavier payloads
Tank-track base	2 motors Rubber tank tracks	Drive over rough terrain	Rough slide-based turning Navigation must use compass sensor Not effective to position a vehicle precisely (skid steering)
Two-legged walker	3 motors	Show off to friends and family	Not able to walk over obstacles Cannot turn easily
Six-legged walker	2 motors	Send pets running for their lives	Not able to walk over obstacles Cannot turn easily

listing chassis function limitations

As with all other design resources, chassis types have limitations specific to each one. For example, if you are using a four-wheeled base without suspension, one drive wheel may come off the ground when the robot travels over a small bump or up a ramp. In some cases, it is better to use three wheels, and, in the case of a balancing bot, two wheels!

design concept

Now that you have a full inventory to work from, you can get down to creating design concepts and put your resources to work.

The four worksheets you have now completed will give you a general overview of the robotic functions you can use and provide a starting point for brainstorming a new design. The following sections will help guide you through that process using the final worksheet.

brainstorming robotic tasks

Instead of trying to determine what kind of robot you can design, it is a good idea instead to see which different robotic tasks can be accomplished by using the functions you detailed in your worksheets. Ideally, you want to match a useful real-world task with the capabilities of the functions in your inventory.

Having a list of functions also helps to identify the information that can potentially be gathered from the robot's environment, or how a robot is able to behave in order to accomplish a repeatable task. It can identify the ways your robot may move and interact with objects and humans.

Once you notice an interesting task that can potentially be accomplished, write down its description in a new Task Worksheet.

If you have trouble getting started on a new idea, try to see if the functions you listed can somehow assist in accomplishing a specific task. For example, the claws or grippers can be used to hold and position an object, and the three-wheeled chassis is capable of accurate navigation and turns. The Rotation Sensor can help with measuring in units such as inches. It is therefore possible to have a robot hold an object accurately in different locations, as demonstrated in Chapter 11's 3D PhotoBot.

Bear in mind that in most cases, you will not be designing a robot that can do everything on its own—don't expect to build a robot that will clean your room without your help. Even autonomous robotic vacuum cleaners get stuck and need a clear path so as not to get in trouble. Your robot will most likely only be able to assist a human in completing real-world tasks. The BenderBot (Chapter 14) and Slot Machine (Chapter 13) projects are good examples of this.

Without going into too much detail, note how your robot will perform each step of the task. Don't worry about the order in which the task steps appear for now, just keep an eye out for limitations. You can also include which special hardware and software functions you think will be required.

Remember there are limitations with every resource, including building parts, so you may want to consider the task's physical and structural requirements.

If you discover a task that may be interesting, the next section will help you decide whether or not the design is feasible.

design feasibility

It is not possible to determine if a design is feasible on paper only—you must build a prototype. However, before physically building a robot, it may be possible to detect issues in how the design is implemented by checking known limitations for the hardware, software, and building parts.

Limitations are provided by the manufacturers, but they can also be logged from your experiences, as well as from information gathered by observing how the parts operate during real-world tasks. You listed some known limitations on your worksheets in the previous sections.

task worksheet (sample)

task description			
Assist a human in taking 3D photos by accurately positioning and moving a digital camera			
key task steps	hardware functions	software functions	known limitations
Accept command from human	Get sound level	Differentiate between different commands (voice recognition)	Microphone not sensitive enough for voice recognition
Move robot around a picture subject	Three-wheeled chassis	Calculate position in degrees Drive in a circle	
Press the camera button hands-free	Get sound level Motor to press button	Morse code–type voice trigger	

There are generally six types of limitations that may affect a design's feasibility.

Manufacturer's specifications   The minimum and maximum thresholds and capacities for a part may be restrictive. Refer to the product's data sheet for specifications.

Observed limitations   You may have noted limitations while testing or observing how the hardware operates over time. Refer to your worksheets.

Availability   You may not have enough parts in your inventory, or parts may already be in use for other functions.

Cost   There may not be a budget to purchase a part.

Technology limitations   The technology does not currently exist or is limited in its functionality.

Knowledge or time limitations   You may have a need to build your own solution but do not have the time or knowledge.

hardware feasibility check

Before you start building a prototype, you may want to ask yourself some questions about how feasible it will be to design your robot and have it perform desired tasks. Although you may have listed a few known limitations for hardware parts, you may now want to take a closer look at each part and ensure that it will work in your design.

For example, ask yourself the following questions:

* Did you check the manufacturer's data sheet?
* Did you check your worksheets for known limitations?
* Are there enough parts in your inventory to build the whole design? If not, can you afford to purchase the required parts?
* If a part is not available, can you design and build it yourself?
* Are the motors fast enough and do they have enough torque?
* Will the chassis be able to carry the payload?
* Can the chassis be reinforced using specific building techniques?
* How large or small does the robot need to be?
* What brightness and noise levels will the robot operate in? Will these levels affect sensor readings?
* Can I use an alternate part not included in the LEGO NXT kit such as a marker, ping-pong ball, dog treat, or NERF gun?
* Does the NXT have enough motor and sensor ports? If not, can I use a Port Multiplexer, which would provide additional sensor or motor ports?
* Can the design be accomplished using more than one NXT communicating over Bluetooth, such as a robot swarm or the Slot Machine project in Chapter 13?

software feasibility check

As with hardware, you need to ensure that the software your design requires will function as expected. For example, not all software functions can be programmed using NXT-G. Some advanced behaviors and calculations need to be written in other languages like C or LabVIEW. If you find that the default NXT-G blocks do not provide sufficient functionality, you may try to locate a custom NXT-G block online.

If you encounter a software limitation and you have programming skills, you may code a solution yourself. Refer to LEGO NXT open source software architecture, which encourages home-brewed creations (see the Extreme section of http://www.mindstorms.com).

If you have trouble locating software, find out if it is available as open source or from the LEGO user community (user groups, websites, blogs).

Don't forget to check the discussion forums for software limitations and to see how others have worked around them.

prototyping and building techniques

Once you have determined a design's feasibility, the next step is to build a prototype. Unfortunately, you don't have building instructions to work from, so a bit of trial and error is called for.

The goal initially is to determine where and how all of the key parts need to be positioned and construct a chassis that will hold them in place using the smallest number of building parts. Just set aside known parts, such as sensors, motors, and special LEGO pieces, and put your LEGO-building skills to work.

There are no set rules for prototyping, but here are some general building tips that may come in handy:

* If your design calls for a drive train, wheels, or motors, start with those.
* Try to make the NXT brick part of the base structure and attach everything else to it.
* Try to position motors and sensors so that the cables don't interfere with the robot's operation.
* Add sensors and manipulators last, but make sure to build in solid attach points for each.

testing and improvements

As you run and test your prototype robot, you will discover new functions, limitations, and potential new tasks it can perform. Don't forget to update your worksheets with new ideas, and don't be discouraged if things break or come apart. Eventually, your testing will lead to a robot that runs more smoothly, perhaps even one that can be rolled out of prototype production!

I read this somewhere on the Internet, supposedly a paraphrase of our old friend da Vinci:

> Life is pretty simple: You do some stuff. Most fails.
> Some works. You do more of what works. If it works
> big, others quickly copy it. Then you do something else.
> The trick is the doing something else.

maintenance and repairs

Now your prototype robot is done, and ready to run. You impress your friends and family, and it operates as advertised for a while, but then to everyone's horror, it fails. All robots break and need regular maintenance, so don't be surprised when something goes wrong. You may need to change your program, adjust a sensor, or move parts around to get things just right.

Not only do the robot's batteries need to be recharged, but if you change its operating environment, the sensors may need to be re-calibrated. You may need to snug up LEGO parts or make structural modifications. If the design fails critically, you can reuse parts of the design that worked well, and go back to prototyping.

the design cycle

The following steps will guide you through the design cycle.

Step 1   Take inventory—determine the design resources that are available.

Step 2   Brainstorm function ideas for resources.

Step 3   Brainstorm robotic task ideas—using worksheets, determine a useful real-world task that could potentially be performed by combining resources in different and new ways.

Step 4   Check the feasibility of the design limitations of resources. Return to step 2 if the design is not possible.

Step 5   Design and build a prototype.

Step 6   Test and improve your prototype design.

Step 7   Repeat and refine your design in steps 2 through 6 until either a working design is completed or it is abandoned due to limitations.

Step 8   Roll out your design—use the robot to perform tasks in a real-world environment.

Step 9   Perform necessary maintenance and repairs.

worksheets

The following pages contain blank worksheets that can be used for the sections in this chapter.

smart parts worksheet

part	qty.	native part functions	software-enhanced part functions	known limitations

smart part groups worksheet

part group	qty.	software-enhanced functions	known limitations

building parts and assemblies worksheet

building part or assembly	mechanical functions	known limitations

chassis types worksheet

chassis type	key resources	functions	known limitations

task worksheet

task description			

key task steps	hardware functions	software functions	known limitations

7

bluetooth on the NXT

When LEGO designed the NXT version of MINDSTORMS, it improved upon the original kit by adding several new features (such as an LCD), improving the motors by adding encoders, and enhancing the sensors (and how the NXT interacts with them). But one of the most exciting new added features was Bluetooth.

With Bluetooth the NXT brick can now communicate wirelessly with other devices, including computers, cell phones, and even other NXT bricks. And with Bluetooth your NXT is no longer alone in a hostile world: It can communicate with other devices around it, which opens up vast new possibilities.

Now to be fair, the RCX brick from the RIS set could communicate with the world around it as well, though via infrared (IR) signals through the deep reddish window on the front of the brick. In fact, IR was used to program the RCX or to download firmware into it, which ultimately caused more than a little trouble in an environment crowded with other LEGO robots—as *anyone* who has worked with FIRST LEGO League (FLL) knows in excruciating detail.

problems with the RCX IR connection

There are at least three main problems with IR communication on the RCX. First, IR's physical nature limits it to its line of sight: If the transmitter and receiver cannot "see" one another, they can't communicate. To download a program to the RCX, the computer and the RCX had to be fairly close together, and the IR tower connected to the computer had to be pointed at the IR window on the RCX. If you lost the line of sight, you lost the connection and had to start all over again.

Second, the IR system was a *broadcast system* with no way to tag a message to go to just one specific robot. The IR system didn't care which RCX-based robot received its data; all robots were equal. As a result, if you tried to download a program to your RCX, and your friend had an RCX-based robot turned on and in the line of sight of the IR tower, both RCXs would try to accept the new program. This would confuse the computer and cause it to stop the process partway through, usually leaving *both* RCXs scrambled.

A third issue was the limited amount of information that could be sent from one RCX to another via IR; a standard RCX "message" was limited to a single byte, just 0 through 255. While you could send complicated information as a long sequence of bytes, one byte at a time, if you needed two RCXs to share a lot of information, it was slow going. Worse, if there were more than two RCXs in view when you sent the message, it was anyone's guess as to which one received the message. This made inter-RCX communication even more challenging.

WHATEVER BECAME OF IR COMMUNICATION?

It may seem that LEGO has completely abandoned IR communication in the move to the NXT platform. While the NXT itself has no built-in IR abilities, LEGO has continued to use IR control for existing products and has actually based one of its new systems (called Power Functions) on a small IR remote control and receiver unit, so IR is far from abandoned. More pertinent to the NXT, there are third-party sensors in development that should act as an "IR bridge," allowing the NXT to use the IR control codes for both the older products (like the RCX, Spybots, and Manas), as well as the newer offerings like the Hobby Trains and Power Functions systems. Far from being a "dead technology," IR may end up being resurrected as yet another way to network your NXT with the outside world.

bluetooth as problem solver

Bluetooth has solved many of the problems with IR. For one thing, *Bluetooth (BT)* uses radio waves (such as Wi-Fi), so the line-of-sight issues go away: It goes through walls and many other common barriers.

Just like a cordless phone in your hand works better than a little IR remote for your TV, so the BT in the NXT easily trumps the IR system for line-of-sight issues. And BT has a clear advantage in range, too: The NXT uses a Class 2 chip, with a range of about 10 meters (roughly 30 feet).

Also, the BT standard requires that devices specifically *pair* with each other. Every BT device has a name or label, which means that, unlike with IR, if you want to send a message to a specific device, you can specify its address and send your message directly to it. When using BT, a program sent to one NXT *cannot* be received by an "innocent bystander" NXT in the same room.

Furthermore, instead of limiting the NXT-G program to sending just a single byte of information per message, you can now send larger numbers (or even text strings or logic values) over BT to another connected NXT.

NOTE Incidentally, there's an important safety feature here. Note that I said "connected NXT." To make a BT connection between two NXT bricks, the connection must be set up manually; there's no way a program can "trick" your NXT into following somebody else's commands unless you first give it permission by setting up a trusted link.

You can think of IR messaging as an analogy for people using semaphore flags or hand signals in a crowded ballroom where you don't know anybody's name: It's confusing, slow, and fraught with the possibility of error. BT gives everybody in the ballroom a cell phone—with a list of names and numbers for everyone else in the ballroom, to boot.

THE TRIUMPH OF EMPIRICISM—BT RANGE

One of the first times I was asked about the range of the NXT's BT communication was in 2006 at National Instruments Week, a major technical conference in Austin, Texas. I stated quite honestly that I didn't know the range. Meanwhile, Steve Hassenplug took a more practical stance and proceeded to drive his BT-controlled robot more than 100 feet away, still in control (and in an environment that had more than a few other BT devices sharing the airwaves, to put it mildly).

making a connection: wireless introductions

Since any BT device needs to have a established connection, the first thing to do with your NXT is to turn on BT and start making some connections. If you haven't done so already, turn on your NXT and navigate to the BT menu (one of the five main menus on the NXT, signified by a funny, stylized *B* in an oval enclosure). Just use the Left and Right arrow buttons to select the BT symbol and hit the orange Enter button to select it.

If this is the first time you've played with BT on your NXT, you will find that BT is off, and you are offered the option to turn it on; do so by pushing the orange Enter button. After it turns on, you have access to several more options in the BT menu on the brick, including the following:

* My Contacts (in which the NXT keeps your wireless address book of previously established connections)
* Connections (if the NXT brick is actively connected to another BT device, it will show up here)
* Visibility (more on this later)
* On/Off (to turn BT, not the entire NXT brick, on or off)
* Search (a way for the NXT to find friends on its own; more on this later)

For now, make sure that Visibility is set to *Visible*. With visibility turned on (the eye is open), your NXT is visible to other BT devices looking for a new contact.

time for a connection

Now, of course, you need something to connect *with*. Here the options become much wider, and there's no way to describe them all in one chapter. Instead, we'll focus on two situations: a BT-equipped computer running NXT-G, and other NXT bricks.

NOTE All computers are not created equal. In fact, it would sometimes seem that no two computer systems on Earth are actually the same. The following section represents a rough overview of how to connect a laptop PowerPC Mac running OS 10.4.8 with built-in BT, as well as a PC running Windows XP Professional either with built-in BT or using an adapter. In fact, user experiences have varied wildly, and we have no way to test (or give directions for) setting up a BT connection for all combinations of computers, BT devices, and operating systems out there, so treat the following as examples, *not* detailed plans.

First, make sure that you have a BT-equipped computer (either built in or using a BT *dongle* that plugs into a USB port, similar to the one LEGO sells) and that the computer has BT turned on, as well.

connecting with a mac

To set up BT on a PowerPC Mac, start by setting up the computer:

1. Start up the NXT-G environment.

2. Open a program. (A new blank program will do, or use one you already have developed; it doesn't matter which.)

3. With a programming sheet open, you can access the *Controller*— the constellation of five buttons in the lower-right corner of the programming sheet. Click the top left one to open the NXT window.

4. With the NXT on and BT visible, click the **Scan** button. NXT-G should now use your computer's BT to scan the surrounding area for BT-visible devices. You might have to select your NXT (still with the default label of *NXT*) from a list of BT devices your computer knows about, as shown in Figure 7-1.

Once you've gone through this discovery process, the NXT-G environment should keep all the information in its internal listing, ready to connect when requested.

If you now try to use the new connection that has appeared in the NXT window onscreen, you should find that the computer has one last hoop for you to jump through: A proper NXT BT connection requires a *passkey*, a simple security code to make sure that both parties approve the connection, and to make this connection secure. Since this is a never-before-used connection, another window should pop up (as shown in Figure 7-2), asking you to "pair" the two BT devices. It even suggests a code to use (*1234* in the case of the NXT). Use a different number if you wish, but just be sure that the *same* number is entered as the passkey on both the computer and the NXT brick.

When you click **Pair** to continue the process and type a code, you'll find your NXT brick beeping for attention, asking for a passkey as well. Press the orange Enter button to use the default passkey (or enter your key if you used a custom one), and the devices will be paired. Your computer should now remember the proper passkey to use for this particular NXT, and (unless you delete all the connection information) you shouldn't have to walk through this setup again.

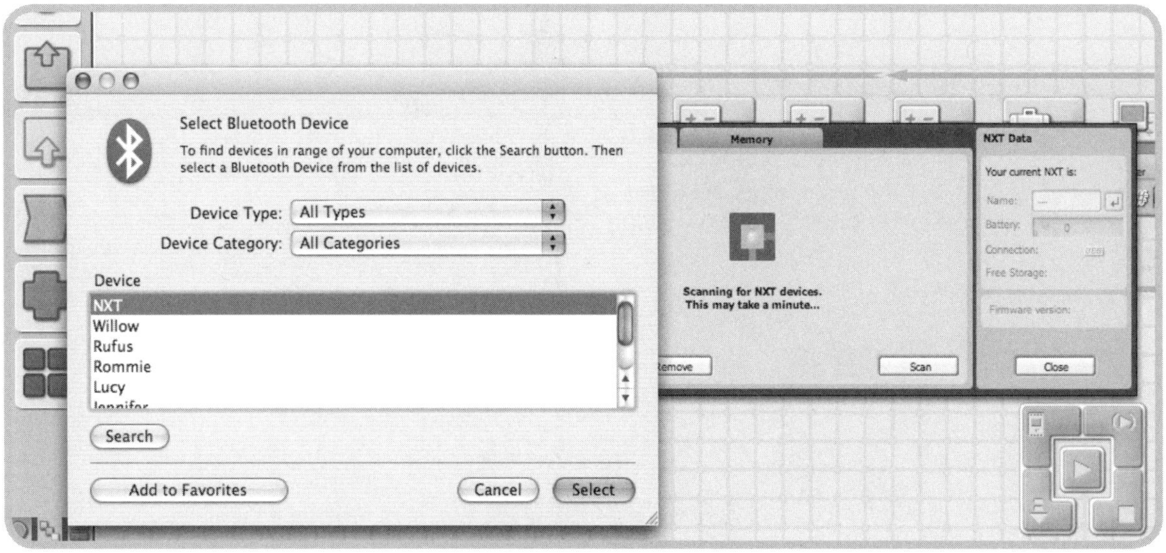

Figure 7-1: The NXT window in the background and the OS X Select Bluetooth Device window in the foreground

DIFFERENT FLAVORS OF BT

Because BT is so universal (connecting BT earphones, computer mice, and even some GPS units), it offers interaction potential well beyond the LEGO system—including, of course, cell phones and PDAs. But not everything uses the same BT standard, so there are some devices (Wii controllers, for instance) that can't directly interface with the NXT.

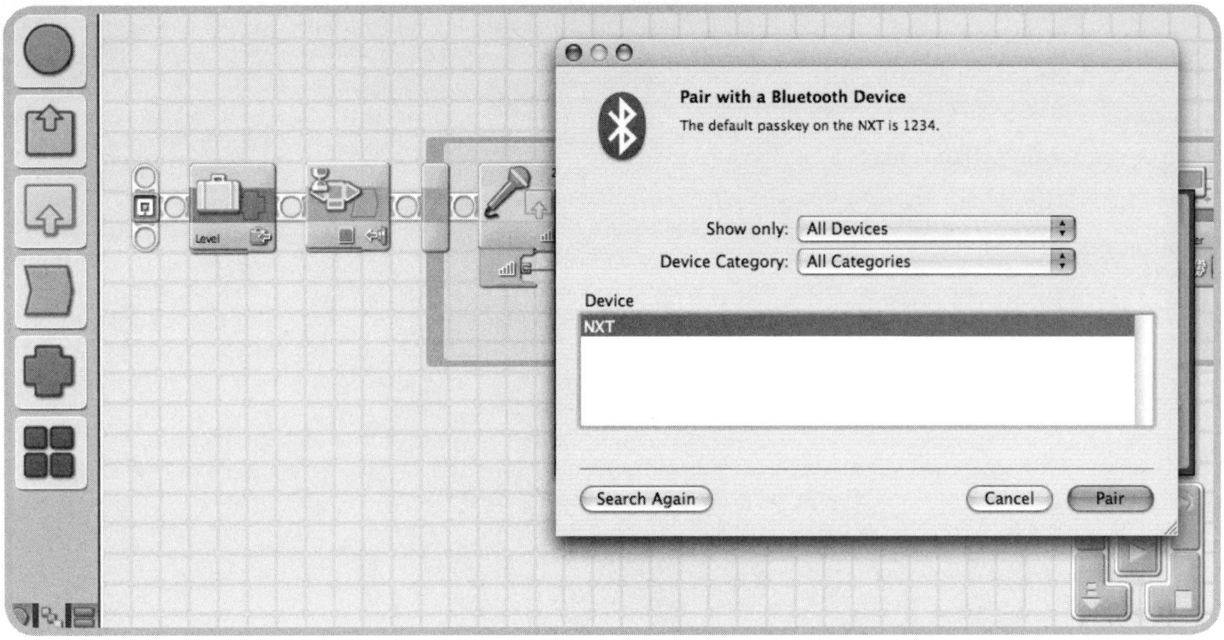

Figure 7-2: The passkey window

connecting with a windows PC

If you're using a version of Windows (such as Windows XP), again, there are many variations for making a connection with the NXT. As an example, Figures 7-3 through 7-9 show how to set up a Windows XP Professional operating system with the BT card currently turned off.

1. Start by enabling the BT card (Figure 7-3).

Figure 7-3: Enabling the BT card

2. Open the My Bluetooth Places folder (Start ▸ All Programs ▸ My Bluetooth Places) and select the **Search for devices in range** task from the left-side toolbar. Figure 7-4 shows that an NXT device called *Jim* is discovered.

Figure 7-4: Finding the BT device Jim

3. Double-clicking the Jim BT device takes you to another screen, as shown in Figure 7-5. The *Dev B on Jim* icon indicates that there is currently no connection. At this point, simply double-click the icon.

Figure 7-5: *The* Dev B on Jim *icon, indicating there is no connection*

After double-clicking the icon, a message lets you know that a connection is being attempted. This is shown in Figure 7-6.

Figure 7-6: *Attempting a BT connection*

4. Your NXT brick will beep, indicating that a BT connection is being attempted. Accept the default security code (1234) by pressing the Enter button; you can also use the Left and Right buttons on the brick to select your own code. If you choose your own code, be sure to remember it, because you will also need to type it in the message window that appears on the PC, as shown in Figure 7-7. After you type the BT security code in this window, click **OK**.

Figure 7-7: *The PC's prompt for the BT security code*

5. If a successful connection is made, a message similar to the one shown in Figure 7-8 will appear. Click **OK**.

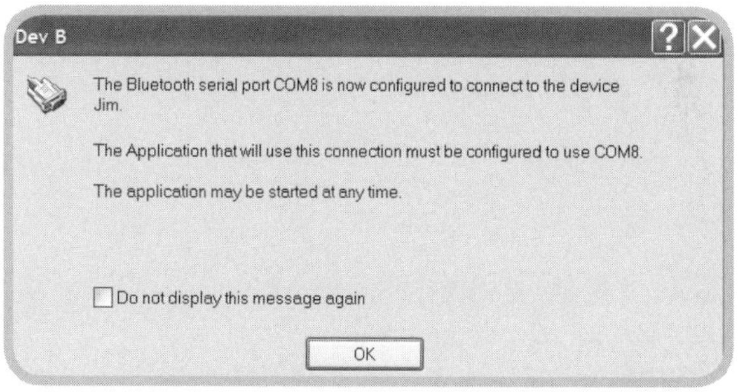

Figure 7-8: *A message confirming the BT connection*

Finally, in the My Bluetooth Places window, you will see the NXT device with double green arrows underneath the icon, indicating a successful connection (see Figure 7-9).

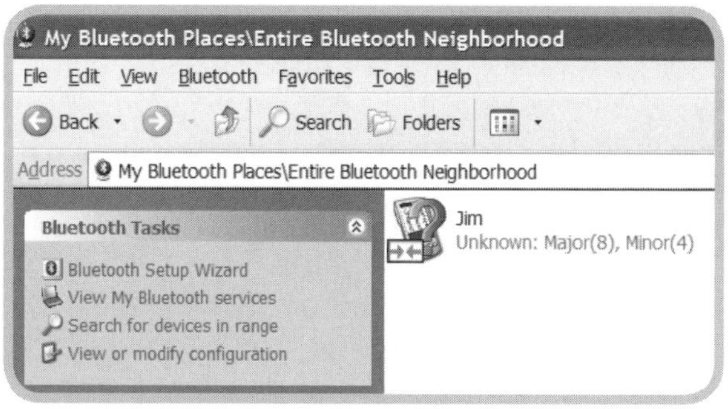

Figure 7-9: *The successfully connected BT device in the My Bluetooth Places window*

after your device is paired

After your NXT brick is connected via BT, the right pane of the NXT window should fill in with all the information from the brick, just as it does when it is connected via the USB cable. The NXT brick should also display its connection state for you: Just to the right of the BT symbol in the upper-left corner, the half-diamond that was there should now show a full diamond, indicating that both sides of the BT connection are active.

Once connected, as shown in Figure 7-10, you can download or upload information, delete files, and so on, just as you would over the USB connection (except that you can only download new firmware with the USB cable). And best of all, the USB leash is gone.

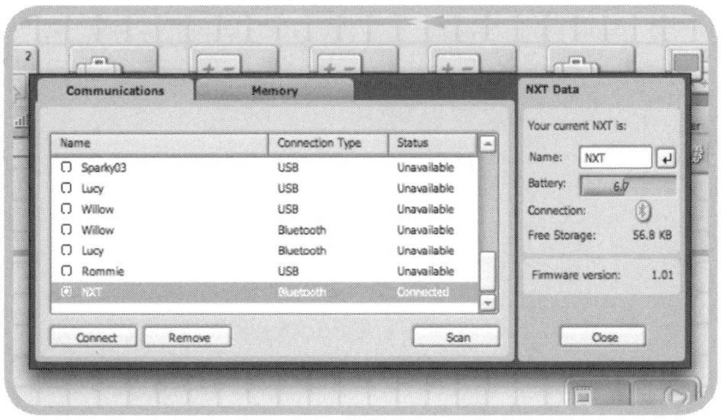

Figure 7-10: The NXT connected via BT

While dispensing with the USB cable is certainly convenient, there are some really practical reasons for using a BT connection. One of the big ones relates to the feedback boxes in the configuration panes of blocks such as the Motor or Rotation Sensor blocks. You might have discovered that these feedback boxes give the current position of the motors (or the current value returned by the sensor, and so on) *if* you are connected to the NXT brick. But it's awfully hard to test or troubleshoot a USB-connected robot using this information, because it is tethered. That USB cable connection keeps pulling it off to the side, or tipping it over, or limiting where it can go. But with a wireless BT connection, you can now view all these dynamically updated feedback boxes in real time, even if the robot is in the other room. A nice trick to use for future troubleshooting.

breaking the BT barrier: troubleshooting

If the preceding sequence worked for you, that's great. But if you had any problems establishing or maintaining a BT connection between the computer and the NXT brick, you are not alone. BT can be tricky.

The first thing to do when troubleshooting the BT connection is to make sure that your BT dongle or computer works with the NXT. While LEGO could not test all the possible combinations of BT dongles, computers, and operating systems out there, it does list certain tested, known-working systems at http://mindstorms.lego .com/Overview/Bluetooth.aspx.

Although we can't troubleshoot every possible issue, we can offer some hints, tips, and general observations to try to ease the potential pain of BT teething. Here goes (this might be a bit technical):

* Some custom drivers that install with the dongle just don't seem to work with the NXT on some systems. Many people have found that on a PC, reverting to the widcomm drivers (instead of custom drivers) sometimes makes things go more smoothly because the communications package that Windows Service Pack 2 installs seems to work well with the NXT.

BT AND FLL: DON'T THROW AWAY THE DONGLE

No, BT is not allowed in FLL competition, but that doesn't mean you shouldn't use it for development. For instance, one team member can work at the table, moving the robot to various positions or positioning the Light Sensor over different targets on the field; while a programmer at the computer can instantly access data from the NXT's various sensors. The programmer can then upload a corrected program to the robot, all without ever having to carry the robot back and forth.

* The Toshiba BT dongles in particular don't seem to work with the NXT, but if you uninstall the Toshiba BT stack and use the Windows Service Pack 2 stack, you might have better luck.
* When in doubt, try uninstalling the BT device and deleting any previous connections listed on the NXT, within the NXT-G environment (in the Connections pane of the NXT window) or within the operating system. Then quit NXT-G completely and restart the connection process from scratch. Or try deleting verified connections on the NXT brick itself and then re-establish them.
* Use the default password (1234) on the NXT for the connection process and answer the computer's request to confirm the passkey as soon as possible. There seem to be issues with the passkey window timing out on the computer side on some systems.
* Make sure to initialize the BT connection from the computer side (not from the NXT brick) and preferably from *within* the NXT-G environment, if you are running a Mac. If the first time you scan to find the NXT it doesn't pop up, don't panic. Rescan to try to find the brick, or turn BT on and off on the NXT brick before rescanning.
* On Intel-based Macs, BT does not currently function under Rosetta. If you have an Intel-based Mac, you need to use a universal binary version of the NXT-G software to have the BT computer-to-NXT functionality. While this binary wasn't available at press time, it might be by the time that you read this.
* Read the manual. For instance, most of the following information is in the paper manual that comes with the NXT.

networking with your peers: NXT-to-NXT communication

If this were all there was to BT on the NXT, it would be nice, but not completely earth-shattering. After all, it seems like a lot of trouble to go through just to rid yourself of one little cable.

One of the really neat things about using BT with the NXT is that it can communicate with other NXT bricks as well, a feature that greatly expands its uses. Using BT, you can network together two (or more!) NXTs, which gives a single creation the chance to use much more than three motors and four sensors.

NOTE While this was possible on the RCX, it was often more trouble than it was worth. The messages that could be exchanged were simple, and the RCXs involved had to have a clear line of sight to each other for the IR to work. And even if all the planets were in alignment, if you weren't careful the RCXs sometimes "missed" a message, or they could be confused by other RCXs using IR messaging.

The IR message system is good (and for some processes, actually still superior), but for most cases the new inter-NXT BT messaging system is better.

You can think of inter-NXT BT messaging as a simple, old-fashioned mail system. Consider three friends who live in separate apartment building and want to exchange mail. To deliver the messages, one friend must agree to be the delivery person (the "postmaster"), while the others end up being more passive partners—only putting their messages out to be delivered. For these messages to reach the correct destination, each friend needs both an address and a specific mailbox at that address to put the message in (remember, they live in apartment buildings, and one building might have multiple mailboxes). But if you think about it, the "postmaster" on delivery duty doesn't need to reveal her own address; after all, she certainly knows how to deliver a message to herself. So the passive partners in this mail system need only label the address *To the postmaster* if they want to send a message to her, while the active partner needs slightly more specific addresses to keep track of where to deliver the messages (she might want to send a message to either one of her two partners). There's just one more twist: While the "postmaster" might be willing to go out now and then and look for messages and deliver her own, she won't just carry the mail from one passive partner to the other. (Hey, she has a life, too—what, you thought all she ever wanted to do was pass mail around for her friends?)

More concretely, say our postmaster is named Amy, while her two friends are Bob and Cindy. If Amy wants to send a message to Bob, she just labels the message with Bob's address and mailbox number, so she'll know where to put it the next time she goes out. If Bob and Cindy want to send a message to Amy, they just label the message *To the postmaster* with an additional mailbox number. But what if Bob wants to send a message to Cindy? Well, he can't—at least not directly, because Amy won't do that sort of "third-person" delivery. If Bob really needs to get a message to Cindy in this system, he has to send a message to Amy (his only connection in this system) and ask her very politely to send a specific message on to Cindy; Bob can't mail Cindy directly. It's a bit odd, but it works. And it has some similarities with the way inter-NXT BT messages are handled in NXT-G.

The point is that there is *no* direct connection between two slave NXTs, but you can send BT messages between two slaves by using the master BT NXT as a relay.

Under NXT-G, messages are directed via connections and mailboxes. A *connection* number is similar to the address, while the *mailbox* number simply determines which of several (10, actually) mailboxes to queue the message into at that address. One NXT (known as the BT *master*) can send messages to any other connected NXT over the "correct" connection, with one connection for each NXT in the network (generally connections [1], [2], or [3]). The other NXTs in the network (termed *slaves*) can send messages only to the master, never to another slave. And since the slave NXTs don't need a special address (because they can send messages only back to the master, not to other slaves), they always send over connection [0]. To further complicate matters, a message going to an NXT can be placed in one of those 10 mailboxes to try to help keep things organized. Figure 7-11 shows a simple "map" of a BT network involving four NXTs.

An NXT is determined to be a master if it initiates a BT connection, and its identity as master cannot be changed while the program is running.

This hints at several important limitations under this system:

* There is no broadcast mode; if an NXT needs to send a message to every NXT in the network, it must do so to only one NXT at a time. (If the sending NXT is a slave, it must send a message "up" to the master, which then echoes the message "down" to the other NXT slaves, one at a time.)
* An NXT can never be both a master and a slave, because a slave NXT has only one connection: back to the BT master in the network.
* If you inspect the NXT-G blocks or the menus on the NXT brick, you'll see only four listed connections, [0] through [3]. This means that under NXT-G, a BT network will interconnect with at most four NXTs: the master with three slaves beneath it.

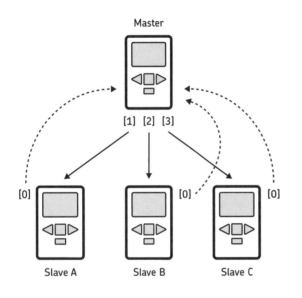

Figure 7-11: A diagram of the specific connections in a typical four-NXT network

setting up an inter-NXT connection

Setting up an inter-NXT connection is simple. To do so, all you need are (at least) two NXTs with BT turned on and visible. To set up a connection for the first time on the NXT that will ultimately be the BT master in this mini-network, do the following:

1. Select **Search** under the BT menu and wait while it tries to discover BT-capable devices in the area. After a while, it should ask you to select the device you want to connect to from a list.

2. Select the NXT you want to network with and then assign it to connection [1], [2], or [3]. (Connection [0] is not available because this is a downlink from master to slave, and connection [0] is reserved for a slave uplink connection.) For this example, use connection **[1]**.

3. If this is the first time these two NXTs have been connected, you have to go through the passkey windows on each to complete the connection. As with the computer, once you have walked through the passkey window to establish a connection for the first time, the NXT will remember it for you. For any future connections you should be able to open the Contacts menu, select the NXT you want to connect to, and then assign a connection to it. It's that easy.

NOTE If you play with BT connections a lot (especially if you are using a NXT-to-PC BT connection to download and test NXT-to-NXT BT programs—as discussed in Chapter 8), you might find that the computer reports that it can't establish a connection. One reason for this is that only one connection is allowed for any BT slave (the one back to the master). Therefore, if you tell the computer to connect to an NXT that is currently configured in an existing network, chances are something will act funny. Either the computer will not be able to establish a connection, or the previous network will be torn down by the NXT being ripped out of it when the computer demands a connection. To avoid confusion, make sure that the device in question is not already locked into some other BT network before trying to establish a BT connection. If it is, disconnect it first. No single NXT can be part of two networks simultaneously.

communicating between NXTs

After your NXTs are connected, you can send information from one NXT to the other, as long as you remember which connection (address) to send (mail) the information to. For example, to give a friend a new sound or program file that is on your NXT, you could set up a BT connection to your friend's NXT, select the file you want to transfer, choose *Send* (denoted by the flying letter icon), and then select the connection over which you want to send the file. The two NXTs should then obediently transfer the file, so your friend ends up with a copy as well. Now you have a way to transfer custom sound files or program files without any need to haul a computer around!

BT messaging under program control

After two or more NXTs are connected, programs running on them can exchange information over the network via NXT-G Send and Receive Message blocks. While the network connection must be set up manually from the NXT menu system, all the message sending and receiving can be handled within a running program.

To send a BT message, drop a Send Message block onto the program sequence beam and configure it. There are three parts to the configuration pane:

* The connection used to send the message out on ([0] for a slave sending a message back to the master NXT; or [1], [2], or [3] for a master NXT sending to one of the three slaves it might be connected to)
* What the message is (you need to specify both its type—Text, Number, or Logic—and its value)
* Which mailbox to place the message in on the receiving NXT

All these parts can be set in the configuration panel, or wired in from some other part of the program through the block's data hub. You can send logic values (*true* or *false*); numbers between –2,147,483,648 and 2,147,483,647 (for those of you with two fingers, that's 2^{32} different numbers, as opposed to the RCX limit of 2^8); or strings up to 58 symbols (like this sentence)—a great step up from the old IR messaging system.

receiving messages

Sending a message isn't very useful unless you have a way to retrieve and read incoming messages. To do this, you'll use the Receive Message block.

The configuration pane for the Receive Message block is simple, with only two things that you need to specify:

* The type of message to expect (and if you want to test the message right there, what to compare it to)
* Which mailbox to look for a message in

But this block isn't very useful by itself. Its main purpose is to get information from the mailbox to pass on to some other part of your program, so you almost have to use the output plugs on the data hub.

When the program sequence hits this block, it looks in the specified mailbox for a waiting message and then tries to compare it with the test value you typed in the configuration pane (or wired in). If the values match, the block puts out a *true* on the Yes/No data plug; otherwise it puts out a *false*.

You can use this feature to control a Switch structure, for instance, branching the sequence depending on whether the correct message has just come in. Even better, the block will output the actual message it received, so you can do further comparisons or analyses of it.

NOTE NXT-G treats a Receive Message block just as it treats any other sensor block: It can use it in a Wait block, as the head of a Switch structure, as the Loop condition, and so on. This capability to embed sensor blocks into these special structures extends to the Receive Message block as well.

Notably absent here is any way to specify which connection to check for an incoming message. In fact, you can't. While this really isn't a big issue with a slave NXT in the network (if a slave gets a message, it knows that it must be from the master NXT because no one else can send message to it), it can be confusing if you are trying to interpret messages coming to the master from multiple slaves. It's like getting a message in your mailbox that says *Meet me at my house*, without a signature or a return address.

Fortunately, there are a couple of ways around this. For instance, each slave can send all its messages as a string of text, where the first part is the name of the NXT sending the message, and the second part is some specific piece of information that needs to be conveyed. Or if all the messages to be sent are numbers (say, codes from 0 to 99), then each slave can send its messages in a different "century": The slave on connection [1] can use numbers from 100 to 199. The first digit (1) indicates where the code came from (the slave on connection [1]), while the next two digits can represent the code or information. The slave on connection [2] can likewise use messages 200 to 299, and so on. Or perhaps simplest of all, each slave NXT can target its messages into a specific mailbox on the master NXT (remember that you have 10 to choose from). The solution you choose (or invent yourself) will depend on the situation and your personal preferences.

NXT-to-NXT remote control

One of the first things that I, Steve Hassenplug, and several other NXT fans did the moment we had two NXTs was to try to use one to control the other. Essentially we wanted to use one NXT as a handheld remote control to send Bluetooth (BT) messages to a second NXT. That second NXT would interpret those messages as commands controlling its motors, thus driving around the room under remote control. The result was far more fun than even we thought it would be: many hours of drive-by-wireless play, smash 'em up derby, and BROVs (Bluetooth Remote Operated Vehicles).

In this chapter, I'll not only show you the finished, remote control solution, but I'll also walk you through some of the design (and debugging) steps along the way—as well as show you some optimizations and bonuses you can include in your own implementations.

Before embarking on this sort of project, you should consider what you actually need. In other words, decide what kind of controls you want and how you might institute them.

choosing a robot platform

A *differential drive* robot is one on which each of two wheels is driven by its own motor. (TriBot is one such example; the robot shown in Figure 8-1 is another.) A differential drive usually has two driven wheels, as well as a single "caster" mounting to ensure that the robot will be stable.

When both motors on a differential drive turn one way (for example, let's say clockwise; it might be different for different robots), the robot rolls forward; when both run in reverse (counterclockwise), the robot moves backward. If one motor turns clockwise and the other counterclockwise, the wheels are driven in opposite directions, and the chassis will spin in place, performing a zero-radius turn (the envy of parallel parkers everywhere).

The differential drive, which is by far the most common chassis type used on LEGO robots, is one that almost everybody has had experience with. That makes it a good target for a remote control. As a significant bonus, a vehicle with differential drive is very easy to control: We just need to be able to control the speed and direction of each motor independently.

Figure 8-1: An example of a simple two-wheel differential drive robot (JennToo)

How about details, such as which tires to use, how much distance there should be between the motors, how to design the structure that holds it all together and mounts the caster, and so on? In other words, where are the building plans?

There aren't any building plans. The reason isn't because I don't have any; this is a software project, so I want to make it as platform independent as possible. By selecting a very common, easy-to-build chassis, I'm making sure that this is as widely applicable as possible, without getting lost in the details of exactly how to build it.

NOTE If you really need a model in mind for this (and you certainly will need one for testing and playing), use TriBot, but *don't* be afraid to try this with your own creations. That's why I'm trying to keep it simple and general.

defining a common language

Now that we know what we're controlling, we need to think about how to represent the commands to send to it.

We know that we need to provide each motor in the robot with a command that tells it how fast to go (set a power level from 0 to 100) and in what direction. We could send a whole series of BT message pairs, one with the power level (a number), closely followed by one with the direction (in NXT-G, that's represented by a logical value, which the Send Message block can also handle). But to cut down on the number of BT messages flying around, we could represent that as a single number between –100 and +100. The magnitude of the number would represent the power to set the motor to, and the sign of the number would represent the motor's direction of rotation. Of course, we need to provide this information for both motors, but now we have a format for the messages, at least.

NOTE I selected this method somewhat arbitrarily, and there are surely many ways to handle this problem. For example, we could have had the remote send two numbers: a signed number from –100 to +100 to represent the power supplied to a Move block that controls both motors at once, and a second signed number to wire in to the steering plug of the Move block. In fact, for some vehicles (such as those that steer more like a car) this approach would probably be much better. How many other "control encoding systems" can you come up with?

defining a control system

Once we know how the robot moves (dual motor differential drive) and have some command formats (signed integers to control motor speed and direction), we need to define a control system to generate those commands. Again there is no single "correct" choice, but I prefer the paddle-like controls that you would find on a tank, with left and right paddles. Each paddle generates commands for one of the motors: If you push the paddle forward from neutral, the corresponding motor should run forward; pull it back past neutral, and the motor should go into reverse.

The speed of the motor should be set by how far you push the paddles: A small deflection translates to a slow speed, while a large deflection would command the motors to run quickly. This action is familiar to us and it gives total control of the motors. It's also an extremely simple mechanical system to build. Each paddle could just be a LEGO studless beam (or beams) attached to the hub of a motor, as in Figure 8-2, with the motor's Rotation Sensor used to sense changes in the control paddle's position. Of course, you could get a lot fancier (like the remote on the right in Figure 8-2), but it's not necessary.

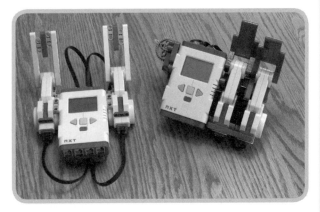

Figure 8-2: Two examples of remote controls: a very simple one on the left and a more refined version on the right

programming the remote control

When designing this robot's control system, we approached it from the robot on up to the remote control with the goal of keeping this project usable on as many robots as possible. But when programming our robot, we'll work from the top down—first programming the remote control to generate the control signals we want. After all, there's no way to really test the receiver program without some set of signals to receive.

To keep the code clean, we'll assume that the control paddles on the remote control NXT brick are attached to motors B and C, and that the wheels on the vehicle are also on those motors.

Basically, the program will check the position of the B paddle, translate it into a BT command, and send that information to the vehicle's NXT brick. If the program can do this continuously with both paddles, the result should be a stream of BT messages that communicate the state of the remote controller to the other NXT. And if we can figure out how to do this for one paddle, all we have to do is repeat a slightly modified version of the same code to read the other paddle. Put this code in a loop and it should work for the remote control.

However, before we get to the loop, we should establish with certainty where the paddles are. To do that, we give the user a message to set the paddles to their "neutral" position; then press and release the orange Enter button. Since this is the default position the user wants for the paddles, the program resets the rotation counters. Now when the user moves a paddle back to the neutral position, the rotation counter will read zero. The initialization completes by clearing the LCD after it's done to give the user some feedback, as shown in Figure 8-3.

To make it easier to remember what is going on in the program (and make the program easier to read), we select all the initialization code and put it into a single My Block. To keep our custom My Blocks straight, we'll give them all the same symbol, according to the text name of the block.

Now that the program knows the neutral position for the paddles, we can start to evaluate the paddle positions and turn those positions into commands to send to the receiver. To do that, we first read the position of motor B with a Rotation Sensor block inside the loop, wiring out an angle (that will correspond to the power to set the receiver motor to) and a direction (the direction flag will indicate whether the paddle was pushed forward or pulled back). We can then use a Switch structure to combine those two pieces of information: If the direction flag is *false*, we multiply the angle by –1 to make it a negative number, while if the direction flag is *true*, we leave it alone (positive). Finally, although we would normally send this out as a BT message, for now we'll just display it on the LCD (for troubleshooting purposes) by wiring out the angle into a variable (call it *scratch*), and then either leaving it alone or multiplying it by –1 to make it negative, depending on the state of the direction flag. Figure 8-4 shows the basic loop that we're using to encode the position of a single paddle.

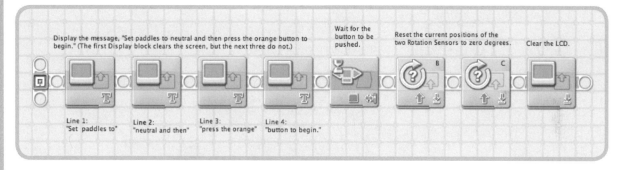

Figure 8-3: Initialization code—printing on the LCD and resetting the paddle encoders

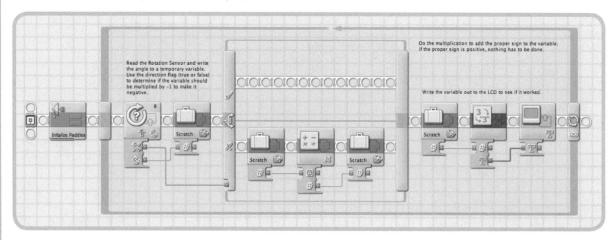

Figure 8-4: Basic single-sensor loop to encode position

Downloading this program to my remote control and moving motor B shows that things are working well: Pushing the paddle one way yields increasingly positive numbers on the LCD, while pushing it in the opposite direction yields a stream of increasingly negative numbers. I don't yet know which way "forward" or "back" will be on this vehicle, so I'll just guess. If I guess wrong, all I have to do is use the opposite signs on the command, which is easy enough to do after the fact.

programming refinements

While this process would work, there are improvements that we can make. For example, if we're careful, we can do away with the variable altogether. Recall from "Matching Wires and Plugs" on page 17 that wires can "branch" only in a downstream direction; that is, two or more wires can't feed into one plug.

Switch structures represent a very important exception to this rule. Because only one state of a Switch structure ever executes at one time, NXT-G *can* actually tell which of the multiple upstream wires is actually carrying information (*valid* in the context

of NXT-G's parent language, LabVIEW) and deliver its contents to whatever lies downstream. It will let you (if you ask nicely) wire multiple outputs from within the Switch to a single input outside the Switch. Knowing this, we could wire the Rotation Sensor angle into the Switch as shown in Figure 8-5, and then wire it out *from both cases of the Switch* to the next block that needs the information downstream.

NOTE Wires feeding out of a Switch statement can be "multi-plexed" behind the scenes by NXT-G. (Thanks go to Kevin Clague for discovering this. Kevin is also very well known for LEGO pneumatics and for having developed the LPub CAD tools for LEGO.)

Inside the Switch, for one case I multiply the angle by –1 (making the resulting angle negative), and in the other case I multiply it by 1 (doing nothing to it . . . but providing a place to start and end the wires). The code then becomes much simpler.

The key to getting this "multiplexing Switch" to work is to be sure that the "edge" or boundary of the Switch is where the wires from the two states join. While it's still true that only one wire can leave the boundary of the Switch, you can have a wire feed into the same point on the edge from each of the cases of the Switch statement.

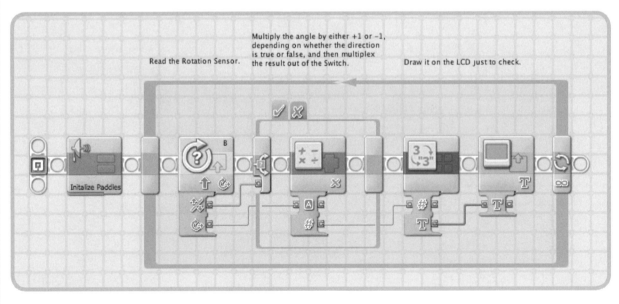

Figure 8-5: Using a multiplexed switch

MULTIPLEXING COMPLEXITIES

Getting this odd multiplexing ability to work correctly on Switches takes some patience, but it can be really useful. Figure 8-6 shows a simple example to walk through, where there are two Math blocks inside each case (true or false) of a Switch structure. For the *true* case, we subtract them in one order, while the opposite order is used in the *false* case. Notice that in step 3, the wires ultimately leading to plug A in the Subtraction block both cross the structure boundary at the exact same point, and only one wire leads from that point to the actual input plug. Once a point on the structure edge is defined this way, you can wire things to it instead of the ultimate destination plug.

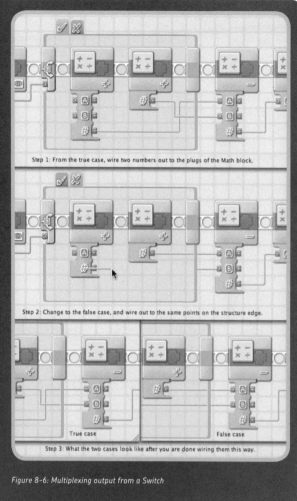

Step 1: From the true case, wire two numbers out to the plugs of the Math block.

Step 2: Change to the false case, and wire out to the same points on the structure edge.

True case   False case

Step 3: What the two cases look like after you are done wiring them this way.

Figure 8-6: Multiplexing output from a Switch

mailing out the result

Now that we've read the Rotation Sensor and turned that reading into a single command, all we have to do is mail it out to the other NXT, using a Send Message block set to send a number. At this point we can also get rid of that display system we used for troubleshooting as well.

We make the handheld remote the BT master, connected to the BT slave (in the vehicle) over connection [1].

NOTE The only reason I selected the handheld unit to be the BT master is that I prefer working the menus on the NXT in my hands instead of crawling around on the floor setting up the BT connection from the vehicle side of the pair.

Since this is the first of two commands (remember, we need one to control motor B and one to control motor A), we mail this command to mailbox #2, and send commands from motor C to mailbox #3 (so that the receiving program knows which commands go to which wheel). For motor C we just duplicate all that code, but change the Rotation Sensor from B to C and the mailbox from #2 to #3.

Once all our code is in the loop, the remote control program should be set. When it runs, it will continue to loop, repeatedly reading the Rotation Sensors and sending that information as commands.

We could stop here, and this program should work, but we can simplify this a bit further by making the code for each motor identical. For example, if we convert the code that reads a single Rotation Sensor (and mails a command) into a My Block, we could use the same My Block for both motors as long as that My Block can set which Rotation Sensor and mailbox to use. To do that, we need to have a way to wire in which motor should be watched and the mailbox to use, of course.

Actually, since we started with mailbox #2, we can simply create a My Block that just needs a single number supplied to it, either 2 or 3. If, for example, we wire in a 2, the My Block should read Rotation Sensor 2 (motor B) and target the outgoing message to mailbox #2. We'll drop a placeholder Math block in front of the code, so we have something to wire to that is outside the proposed My Block, as shown in Figure 8-7.

After we create that My Block, we can delete the placeholder Math block (we only used it to create the wire into the My Block). Next, we'll drop a second copy of that My Block (which I've called Send Command) and configure the first copy to handle port B (by typing a 2 into the port field in the My Block's configuration pane) and the second copy to handled port C (a 3 in the port field). The resulting program looks almost trivially simple, and is significantly smaller (memory-wise) than if we hadn't used My Blocks, as shown in Figure 8-8.

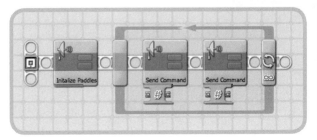

Figure 8-8: A very compact, self-explanatory remote control program

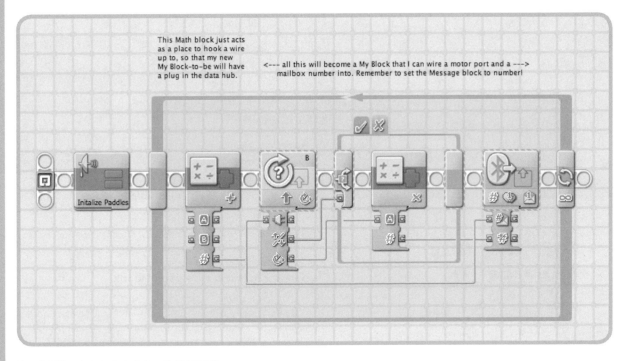

This Math block just acts as a place to hook a wire up to, so that my new My Block-to-be will have a plug in the data hub.

<--- all this will become a My Block that I can wire a motor port and a ---> mailbox number into. Remember to set the Message block to number!

Figure 8-7: The code, just prior to ripping a My Block from it

I use a lot of My Blocks, often with multiple plugs and nested several layers deep, which makes it tough to remember what to wire into each particular plug. To solve this problem, try renaming the plugs using the Comment tool. For instance, when I created the Send Command My Block, I immediately double-clicked it to open it up, and renamed the plug *Port*—a much more descriptive label. This label will now pop up as a tooltip if I hover the mouse over that plug in the data hub, as shown in Figure 8-9.

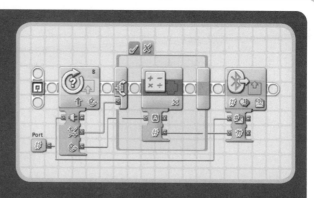

Figure 8-9: The Send Command My Block with the plug relabeled

the receiver

Now that we have a program in place to transmit movement commands, we can move on to the other half of the system: a program that will receive and interpret those commands to move the vehicle.

Again, we want a loop here because the receiver will need to repeatedly convert incoming messages into power and direction commands; then wire those into a Motor block. Ultimately, both motors have to be handled this way.

Our first thought might be to use a Wait for Message block. But the problem with that approach is that although that block would wait for a message, it would wait for only a certain message, and leave us with no way to get that message out for further processing. A better approach is to use a Receive Message block, which would get a command from the mailbox and compare it to zero. If the result of the comparison is greater than zero, the logical output of the Compare block could be used to control a Switch structure, with the *true* case set to run the Motor forward with the power (determined by the value of the command) wired in. If the result of the compare is less than zero, the *false* case would be selected, with the Motor block set to run in reverse, with the power again determined by the value of the command (something like Figure 8-10).

NOTE If you wire in either a positive or a negative number into the Motor block's power plug, it will treat it as a positive number, ignoring the sign.

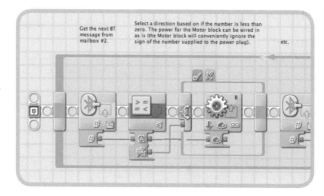

Figure 8-10: Decoding commands for motor B

This shows how to control motor B, but we need a very similar piece of code to control motor C as well (you can just see a hint of it on the right edge of the figure). We set the Motor block to a duration of *unlimited* so that the motor will run continuously at whatever power level it was last set to until it receives a new command.

Now, while the above method should work, we can make it even cleaner. Notice that when we wire a direction into a Motor block, we're just wiring in a *true* or *false*, which is exactly what the Compare block wires out. Although in the program above we used a Compare block to select which case of the Switch structure to execute, we could simply wire the logical wire straight into the direction plug of the Motor block, as shown in Figure 8-11.

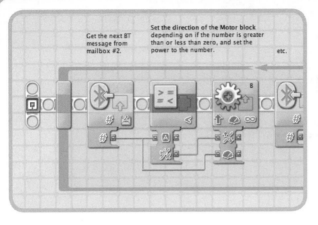

Figure 8-11: A simpler motor control scheme

control issues

Well, at least in theory, anyway. While that seems like it would work, if you try it you'll find it doesn't. Instead of moving smoothly, the robot moves in a series of very short jerks, as if something is repeatedly stopping the motors almost as soon as they are started. Since the Motor blocks in the program are set to an unlimited duration, something very odd must be going on.

The answer lies not with the motor control, but with the way BT messages and mailboxes work. Think about playing a game of chess through the mail. You arrange with your opponent to send you one move every other day. You go to your mailbox on Monday and your opponent's move is there, so you make your move and continue the game. But on Wednesday, when the next move is supposed to come in, you find your mailbox empty. Has your opponent forfeited her turn, and given you a "free" double turn? Probably not, and it would be a mistake to interpret it that way. The mail might simply be delayed, and you should interpret the lack of a message as a message saying "I'm doing nothing."

Now consider the NXT-G mailbox system. There are 10 mailboxes, and each mailbox can hold 5 separate messages. The NXT stores the messages as it gets them, in case its program doesn't have time to deal with them when they are received. The NXT doles out the messages each time you "look" in the mailbox, using the Receive Message or Wait for Message blocks. But if the program running on the NXT tries to get the next message in the mailbox and finds the mailbox empty, there's a problem. *Something* will still come out of the data plug on the wire, but whatever is put out onto that wire has no connection to a received message. Instead of a valid message sent from the other NXT, what comes out on the wire is often just a zero. In the case of our receiver program, this zero will be interpreted as a motor power of zero, which stops the motor.

When the program finds the mailbox empty, we need to give it a way to detect that and ignore any "junk messages" that might pop out of the wire.

NOTE Why not have the Receive Message block hold up execution of the sequence until a "good" message comes in? Because sometimes you don't know when the next message will arrive, and you might end up delaying the program for a very long time. With this mechanism you have control over the program execution.

NXT-G provides a solution to the empty mailbox problem in the form of a special plug labeled *Message Received* in the data hub of the Receive Message block. When there are pending messages in the mailbox selected, this plug outputs a *true*, but if the mailbox is empty, it outputs a *false*. Since we want only valid messages, we need to keep checking the mailbox until this Message Received flag is set to *true*; only then do we know for certain that the message that was wired out of the block is valid.

To use this block, we simply drop it into a loop and have the loop cycle around and around until the plug outputs a *true*. Once the output is *true*, we know that the message last wired out of the block is valid, and we can process it. So, in place of the simple Receive Message block in the previous version of our program, we'll use this loop structure as essentially our own "wait for a valid message" block, shown at the head of the main loop in Figure 8-12.

With this change, the two programs work together quite well: RC.rbt reads the positions of two of the motors and sends this information as formatted commands (via BT) to the receiver.rbt program. The receiver.rbt program, in turn, interprets those commands as motor direction and power settings and then applies them.

REGARDING MAILBOX OVERFLOWS

In addition to concerning ourselves with empty mailboxes, we also need to take into account potential mailbox overflows. Specifically, if for some reason the mailbox is full and a sixth message comes in, the oldest message held in the queue will be deleted. While that's fine for this particular program (we always want the freshest commands, anyway), we need to be very careful about this potential problem when two NXTs are exchanging a sequence of information. Returning to the chess-by-mail analogy, it's not a very good game if now and then you "miss" the opponent's move.

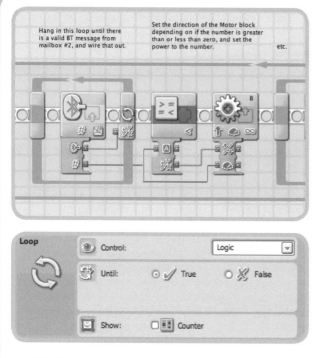

Figure 8-12: The front part of the receiver program with the Loop structure inserted, and the Loop configuration panel shown below

tuning the program

And now a final adjustment. While ideally, we'd like the robot to go forward when we push the paddles forward, it might not be the case when we first try it out. To get the behavior we really want, we might have to switch directions on the motor. Instead of doing a bunch of reprogramming, we simply change one setting in the Comparison blocks in the receiver program.

As initially programmed, any negative command (that is, less than zero) would generate a *true* out of the Compare block. But if a negative command results in your robot going backward, just change the Compare block from *less than* to *greater than*. As a result, commands that previously generated a *true* will generate a *false*, and vice versa. That's all it takes to get things moving forward again.

Similarly, we need to tune the amount of travel for the control paddle on the remote. If, for example, the paddles can be pushed forward only 30 degrees, there would be no way to command the vehicle's motors to 100 percent power, because the magnitude of the command sent is equal to the angle in this system. But that doesn't have to be the case. You could simply multiply each

command by 3 before wiring it into the Motor blocks, thus scaling up the power. That way when you push your paddles forward 30 degrees, and the command "+30" is sent to the vehicle, the program will interpret this command as a power level of 90. The result would be something that is much closer to the full range of available power settings.

where do we go from here?

There are a number of ways to improve this program. For example, you could:

* Make the receiver program more memory efficient, as we did with the remote control transmitter program, by using My Blocks again.
* Reprogram the system to handle all three motors, instead of just two. (In fact, doing so for the transmitter program would require the addition of a single Send Command My Block, configured to port 1. See how My Blocks make things easier?)

I'm sure, too, that you can think of many more interesting robots to apply this program to than a simple TriBot. But first, an important point, as first noted by my wife when she saw me laughing giddily as I drove my little robot around the living room. In her words:

"You just turned $500 worth of hardware into a simulation of a $40 radio-controlled car?"

Though I hated to admit it, she had a point. As much fun as this is, there are certainly cheaper ways to create a remote-controlled robot, including a number of applications that will enable you to command the NXT directly from a cell phone, PDA, or other such device. So what's the point? Why bother to build this fancy radio-controlled vehicle?

Well, for one reason, unlike most remote-control options, both the remote control and the receiver in this setup are functional computers, not just simple mechanisms. They can inspect sensor readings, calculate and display information, and even take autonomous action, if necessary. Neither partner in this system is "dumb"; both can be programmed to do some complex operations on their own.

For instance, the vehicle might work under remote control until it sensed (via the Ultrasonic Sensor or some other mechanism) that it was being driven into a wall, at which point it might "decide" on its own to stop and reverse course to protect itself. Or it could detect when its battery was low and prevent you from driving it quickly, even ignoring (or even complaining?) if you exceeded some power limit.

The remote control is, of course, also a fully functional robot, so it can do things on its own, such as moving the paddles (remember, they're hooked up to motors) forcefully toward neutral if the vehicle takes control of itself.

giving control back to the robot

As an example, let's modify the system so that when you push the orange Enter button, the vehicle will drive straight forward on its own until the Ultrasonic Sensor detects an obstacle in front of it. Of course, this requires a Ultrasonic Sensor on the front of the vehicle, but that shouldn't be too hard to mount. To make the programming easier to explain, I'll assume you have plugged this forward-facing Ultrasonic Sensor into input port 4.

First we'll need to program the remote control to handle this new autonomous behavior. When the user presses the orange button, the remote control needs to do two things: send a message to the robot telling it to enter autonomous mode, and give the user some feedback.

After sending out movement commands, the remote control checks the orange button at the head of a Switch. If it is not pressed, it just prints *Manual mode* on the LCD and continues, letting the sequence loop back and send another set of commands to the vehicle. But if the button is pressed, it sends out a special command, "Go auto," to mailbox #4 on the vehicle. Next, the sequence prints *Auto mode* on the LCD to tell the user what's going on and then waits for the text message "Manual" to come back from the vehicle. Once it receives this message, it knows that the vehicle has finished the autonomous behavior and it can resume sending "normal" commands for manual control (see Figure 8-13).

I put this Switch on Button structure right after the Send Message My Blocks in the original program, but it could go anywhere in the loop, as long as it is checked frequently to be sure that any presses of the orange button are detected before the button is released. In the "pressed" case of the Switch, we use a Wait for Message block to hold up the program until a specific message comes in. I'm also using mailbox #4 instead of the mailboxes that I was previously sending the commands to in order to make it easy to watch and interpret these "special" messages in a different way from the normal motor control codes. If I have 10 mailboxes, I might as well use them to keep things working smoothly.

adding brains to the vehicle

On the vehicle side of things, our code needs to check mailbox #4 frequently to see whether the special command "Go auto" has been received. We can put this code right inline at the end of the loop; again, when it's checked isn't as important as that it be checked frequently. But I don't want to do this the same way as for the regular commands. Remember that they are always coming in, so we had the sequence hang in a loop until a valid motor command came in . . . since we were certain there would be another command coming in very soon, this behavior would not hold up the code for long. But the "Go auto" command may be sent only very infrequently, and if we had the program wait until it arrived, it would get stuck in that Wait block, never checking for new motor commands. The result would be a truly unresponsive robot that accepts the two motor commands and then appears to hang, not responding to the remote control until you hit the orange button to get it out of the waiting state it's locked into.

Instead, we'll have mailbox #4 checked by a Switch on Message. The idea is to have the Switch check for a specific message (in this case a text message) in a certain mailbox. If it is present, the sequence follows the upper branch; otherwise it takes the lower sequence (see Figure 8-14).

When autonomous mode is triggered, the robot beeps to give the user some feedback. Next, it uses a Move block for motors B and C, set to a duration of unlimited to roll the vehicle forward, but not delay the execution of the program sequence by waiting for it to complete some specific distance. The third block just waits for the Ultrasonic Sensor on port 4 to return a distance of less than 25 cm. Finally, the vehicle sends a message (the text string "Manual") back to the remote control to tell it that the autonomous mode is completed. When the remote control unit gets this command, it will start looping again and send out motor commands, and the vehicle will again be ready to receive them.

There's another important point here: that last Send Message block in the above sequence is supposed to send a message from the slave back to the master. Remember that while I established the BT connection as connection [1], that's the connection the master uses to send to the slave. For messages going from the slave to the master, the slave always sends on connection [0]. For a slave, there is only one valid connection, and it is always connection [0].

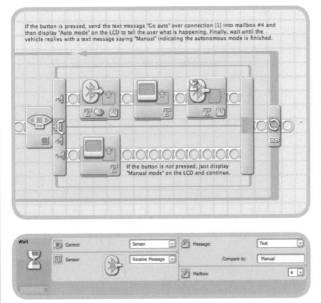

Figure 8-13: Sending out a command to go to auto mode

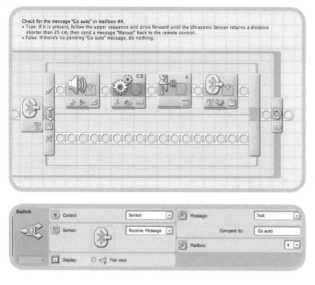

Check for the message "Go auto" in mailbox #4.
• True: If it is present, follow the upper sequence and drive forward until the Ultrasonic Sensor returns a distance shorter than 25 cm, then send a message "Manual" back to the remote control.
• False: If there's no pending "Go auto" message, do nothing.

Figure 8-14: Putting some autonomous behavior into the vehicle

NOTE If a slave NXT can send messages only over connection [0], why are there other available options? Why doesn't the Send Message block set the connection to send over to [0] and simply prevent the selection of other options? The reason is simple: The NXT-G environment has no way of knowing whether the code you are writing is supposed to run on a master or a slave NXT, so it can't make those decisions for you. It is up to you to keep the connection scheme straight!

future directions: going beyond the book

I've developed a reasonable little remote control program here that will enable you to use one NXT and two motors to control a second NXT vehicle. In the course of developing this program, I've demonstrated the Send Message block as well as three different ways to receive messages: the Receive Message block, Wait for Message, and Switch on Message. I've also demonstrated two-way communication.

But this is nowhere near all that you could do, and I encourage you to take this code and these ideas and modify them as you see fit. Here's a very brief list of ideas to get you started:

* Make the autonomous behavior much more complex, including backing up, following a line, or just about anything else. Have the robot remain in autonomous mode until the user pushes the orange button on the remote control a second time.
* Add a third motor to work a claw or other appendage, and have a third motor or control paddle on the remote control it.
* Use BT communication to remote control a robot with rack-and-pinion steering, with one motor driving it forward and backward and the other motor controlling the steering angle of the front wheels.

RUNNING A POST OFFICE IS HARD WORK

A word of warning regarding two-way BT messaging is appropriate at this point. Recall that when using BT with the NXT, the master does all the message shuffling, including polling the slave now and then for any messages waiting to be sent back to the master. All this interrogation and response takes time, and since it's all being executed by the master NXT, things can slow down and get confusing if there are a lot of BT messages streaming rapidly back and forth. If you put a lot of this sort of communication into an application such as a remote-controlled vehicle, you might find that control slows or becomes less responsive. You can try to work around some of these issues by making sure there is sufficient time for everything to happen, but understand that this communication process is a rather processor-intensive one for NXT-G. Therefore, it's usually a good idea to minimize the number of BT messages being sent.

* Give your vehicle a "horn," so that when you push a button on the remote control the vehicle beeps or even speaks a short phrase using a sound file.
* Have the vehicle send information back to the remote control to display on the LCD for the user, such as the distance to the nearest obstruction, how fast it is going, the ambient noise level in the area, and so on.
* Record the movements that the user puts the vehicle through, and then (upon command) try to reverse those motions to "return home."
* Have one remote control more than one NXT vehicle, either simultaneously (sending the same motor commands out over two different connections) or one at a time (you could use the buttons to select which NXT to control). Try using multiple vehicles to perform more complex tasks.

And the possibilities for BT messaging on the NXT are not at all limited to remote controlling a vehicle. For instance, you could:

* Play cooperative games among as many as four players, each with his or her own NXT.
* Text-message friends on an NXT-work (with your own personal short-range pager).
* Use up to four autonomous robots to coordinate on a task or set of tasks.
* Coordinate multiple NXTs in single robots.

The latter is actually one of the biggest advantages of BT communication on the NXT. Not only does that mean more processing power for complex tasks but also (perhaps even more importantly) more sensors and motors built into one robot. By networking together four NXTs (one master communicating with three slaves), a single robot could command 12 motors and 16 different sensors! Truly, with BT communications, the gloves are off!

LIMITATIONS OF BT NETWORKS

You might notice I keep mentioning "as many as four" robots or something similar in a BT network. Why can't you have more? The answer lies with the specific chip that is handling most of the BT networking. While there are BT chips (Bluecores) that can handle more BT connections than three slaves, they are more expensive than the one used in the NXT. The decision-makers at LEGO guessed that most folks would not have that many NXT sets laying about to network (which is probably true for most folks) so they chose the less expensive option. Nonetheless, love it or hate it, the "no more than four" limit is currently built into the hardware itself, and cannot be changed. There might be ways around it (for instance, if an NXT could be programmed to create and tear down BT connections under software control, forming a dynamic network), but for now it is a limitation of the system that you should be aware of.

PART II:

the
robots

RaSPy: a rock, scissors, paper-playing robot

This book had an interesting starting point. Back in May 2006, three of The NXT STEP blog contributors tried an experiment—how difficult would it be to create a robot, program it, and document it if all of the contributors were separated by great distances? Brian Davis (Elkhart, Indiana, USA), Matthias Paul Scholz (Freiburg, Germany), and Jim Kelly (Atlanta, Georgia, USA) used email and the Internet to share pictures, files, programs, and discussion to take a simple idea and turn it into a working robot (see Figure 9-1).

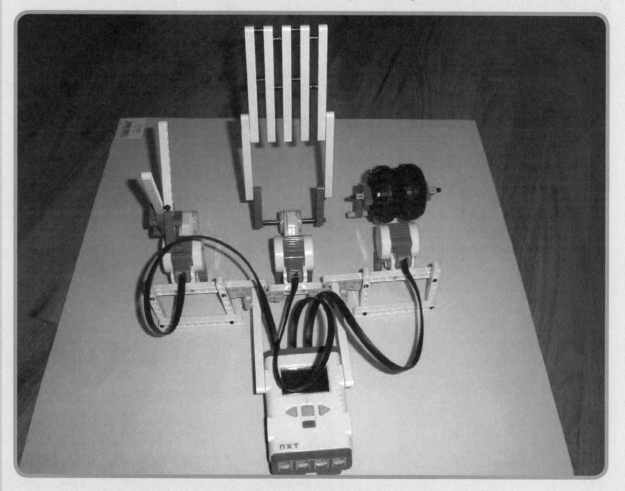

Figure 9-1: An early prototype of RaSPy

That robot is called *RaSPy*, and he plays the simple game of Rock, Scissors, Paper. (Yes, some people call the game *Rock, Paper, Scissors*, and *RaPSy* is just as good a name, but RaSPy just stuck.)

Jim did most of the robot construction. In Figure 9-1 you'll see an early version of RaSPy. The original was *huge* and very unstable and wobbly. Through experimentation, RaSPy (version 2) was created and proved to be much more stable.

Brian took on the task of programming RaSPy, and the three contributors tested and debugged the small robot until it worked perfectly.

The final step was creating the building instructions. Matthias took on this responsibility and created the instructions you'll find in this chapter.

What did the three contributors learn about working together via email and filesharing? Well, the first thing was that testing and modifying a robot can take a long time. For example, all three contributors were in different time zones, which caused some delay in email and discussions. Another important lesson learned was

that sometimes having additional designers can help refine certain rough spots in a robot. When one team member would get stuck or encounter a problem (either with building or programming), another member of the team would often have a solution.

We're including RaSPy in this book because the idea for the book really started with this small robot. As The NXT STEP blog team grew in size, it became apparent that the members were able to provide assistance to one another when it came to designing and programming robots. Many of the building instructions for the robots in this book were done by members of the team who had experience with the software required to create the instructions.

Therefore, you can understand why we're proud of RaSPy—this little robot helped us develop an idea and turn it into the book you hold in your hands. And now, it's your turn. We hope you'll take the building and programming instructions for RaSPy and build the little fellow, modify him, tweak him, change him, and maybe come up with your own version. Have fun!

building RaSPy

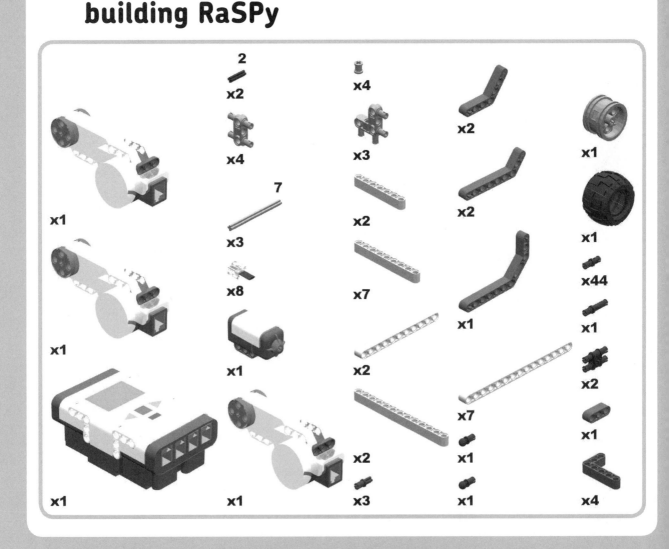

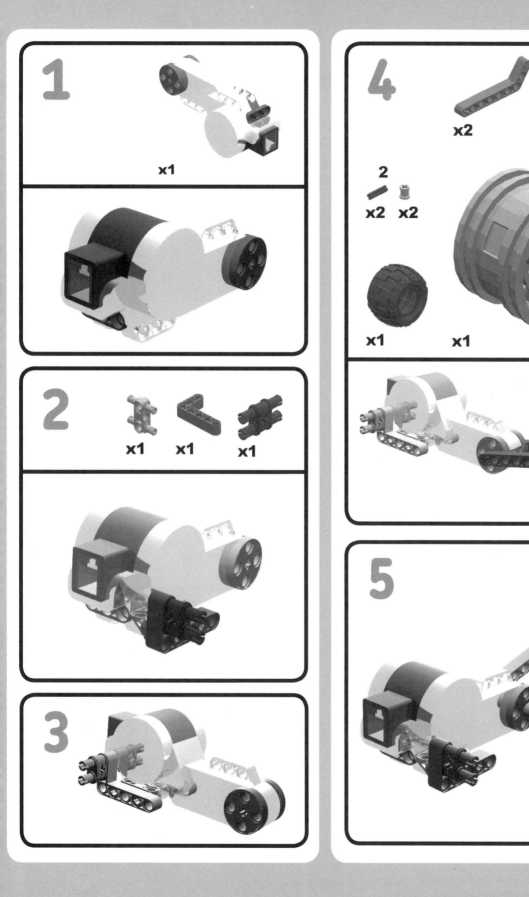

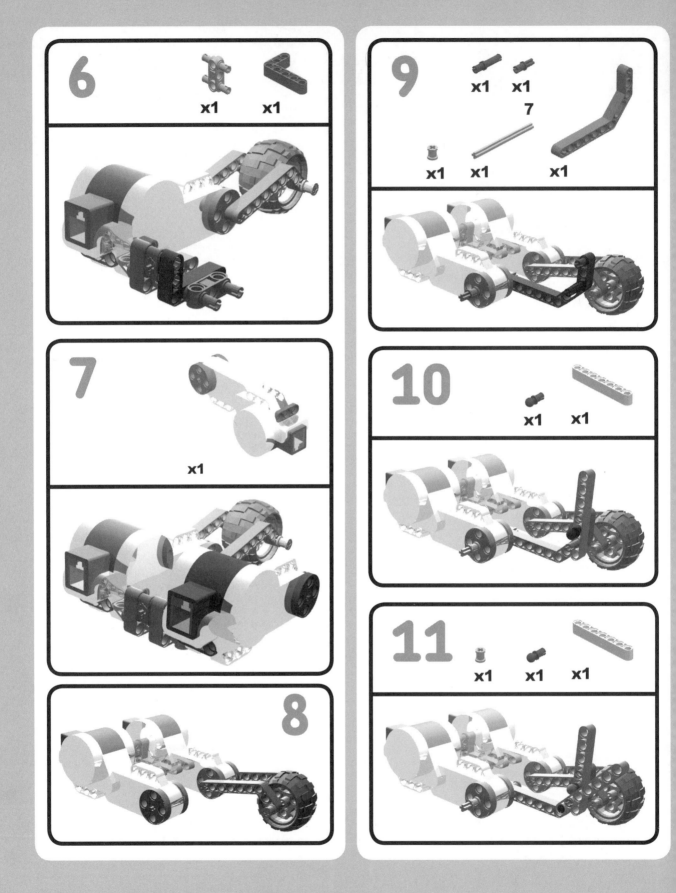

12

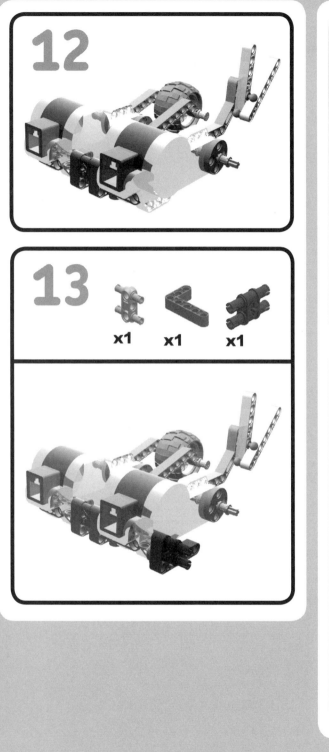

13

x1 x1 x1

14

x1 x1

15

x1

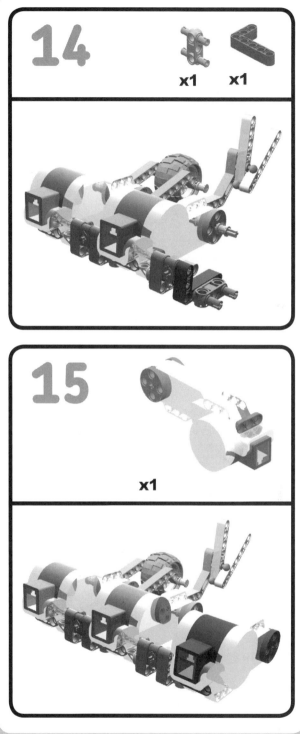

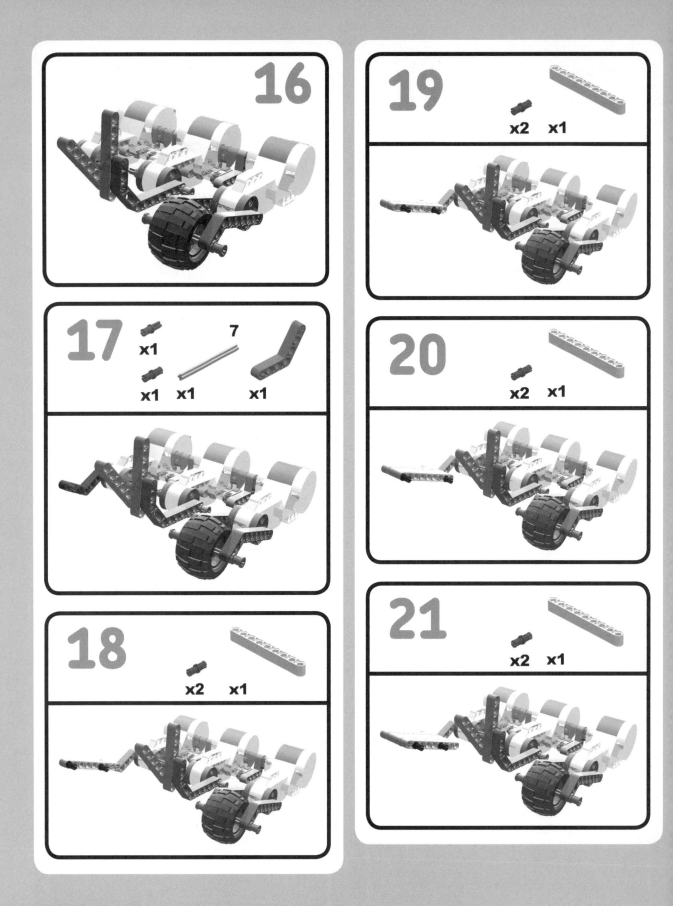

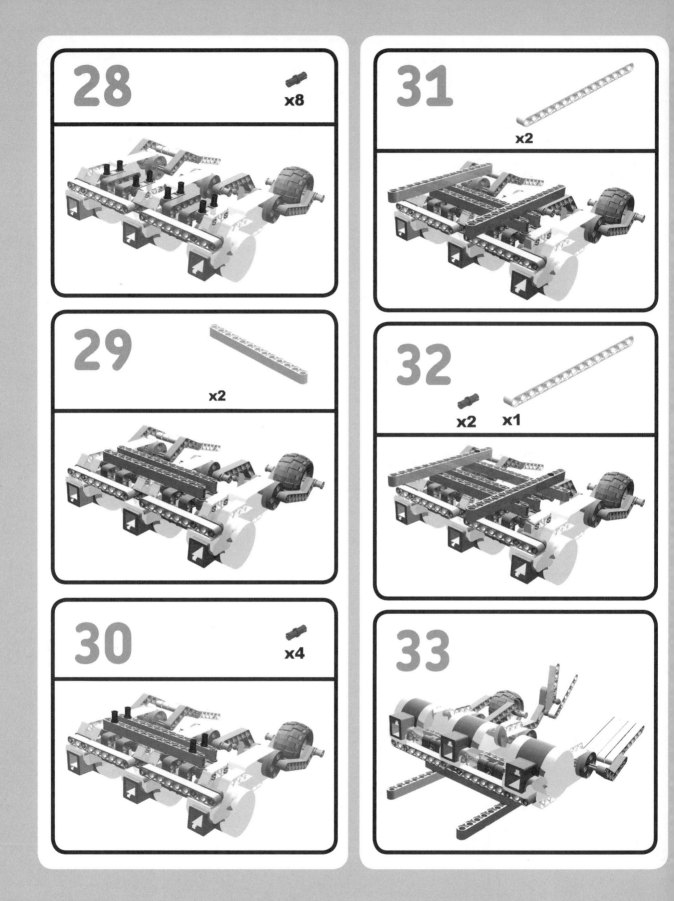

34

x4

35

x2

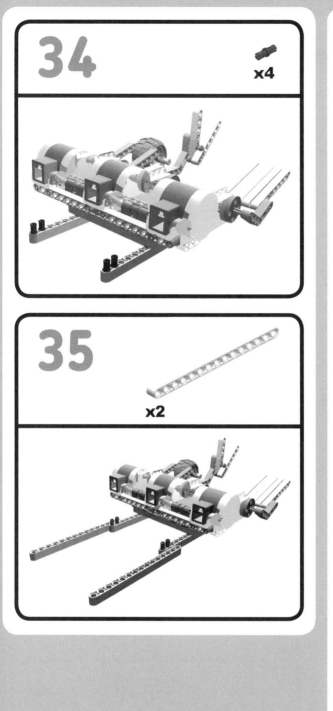

36

x8

37

x1

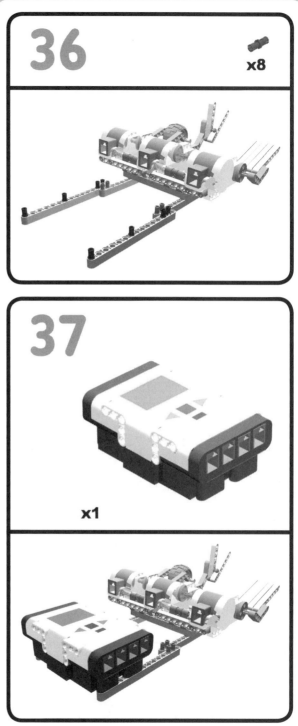

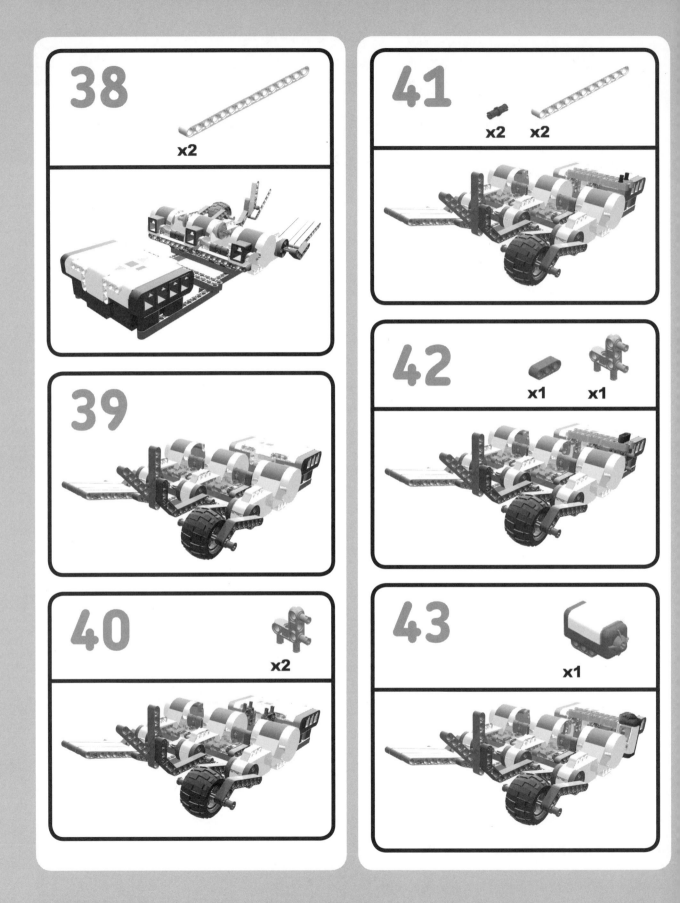

programming RaSPy

In order to program RaSPy to play Rock, Scissors, Paper, we need to figure out exactly what he needs to do. This isn't quite as difficult as teaching a human how to play—after all, a human would need to understand the rules, judge who won each round, and so on. All we care about is that RaSPy can imitate the steps involved in playing a game of Rock, Scissors, Paper. Let's think it through.

Every round begins when the players agree to start, and there is a sequence of three rhythmic *beats*, or hand motions, that synchronize the players before they *throw*, or reveal their choice, on the fourth beat. If you had to list that in detail, you could do it this way:

1. Start the round.

2. Make three motions to set up the timing for the players.

3. On the fourth beat, reveal the symbol you selected.

However, you can see that there's at least one thing I unintentionally left out—actually *making* the choice. This would obviously have to come before step 3. Additionally, this list only has each step happen once. That's okay, but it's not ultimately what we want. To play a game of multiple rounds, it would be nice if the robot immediately got ready for the next round and set itself up to repeat the actions. In NXT-G (and most other programming languages), that's what a loop is for. By realizing this in advance, we know to drop the loop structure *first* and build the other steps within that loop. Sometimes a little bit of planning at the start can save you a good bit of time and effort later on. Here's a revised list of actions for the robot to take:

1. Start the round.

2. Make three motions to set up the timing.

3. Select which symbol to throw.

4. On the fourth beat, indicate the selected symbol.

5. Do it all again (go back to step 1).

Let's create the overall loop structure first, since it works best that way in NXT-G. Drag a Loop block onto the programming sheet as the first block in the program (Figure 9-2). With that in place, we can start working on the above steps in the program.

Let's start the round off with the human player pushing the Touch Sensor. Since the program needs to wait for this event to occur, just use a Wait block set to wait for a Touch Sensor (on port 1, for RaSPy) to be "bumped" (also shown in Figure 9-2). That means the player will have to push *and release* the Touch Sensor before the program sequence continues.

The next step would be to have RaSPy make three motions. Jim originally wanted this to be all three of RaSPy's "hands" moving down and back up, similar to how a human player's hand moves in the game. Clearly, this is another place we could use a loop, as this is a very repetitive motion. As the second block within our main loop, drop a second Loop. However, we only want this loop to cycle three times (once for each of the three beats), so we need to limit it somehow. We can do that by changing the Inner Loop Configuration panel so that the Control is by Count, and while we're at it, set Until to 3. Also, check the box next to the word *Counter* so that the loop shows its counter (Figure 9-2). (Watch the left side of the loop when you do this; you'll see a little plug pop up there.) Now we have a loop within a loop.

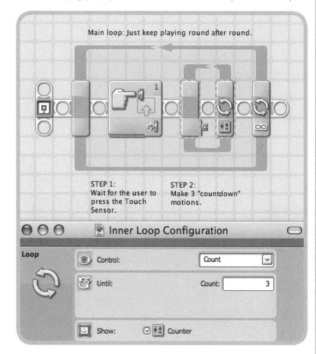

Figure 9-2: Using the Inner Loop Configuration panel to allow RaSPy to count to three

We could continue working on step 2, but for now let's leave it as an empty loop—just a placeholder for the code to come. Instead, let's work out steps 3 and 4, as they are fairly simple.

The program needs to select one of three options (rock, scissors, or paper), but it needs to do it randomly. Right after the inner loop, the program should generate a random number: a 1, 2, or 3, generated by an appropriately set Random block, as shown in Figure 9-3 (with the minimum set to 1 and the maximum set to 3). Step 3 is done!

Now, in order for RaSPy to signal his choice to his human opponent, step 4, he needs to move one of the three motors to throw his chosen symbol. An obvious way to do this would be to wire the output of the Random block into a Switch structure, where all three possibilities could be handled . . . but there's a much easier way. Each of the motors can be referenced by its port number, 1, 2, or 3—the same numbers the Random block generates. As you can see in Figure 9-3, all we need to do is wire the

output of the Random block into the "port" plug of a Motor block, and then have that Motor block rotate the selected arm down a certain number of degrees (setting the duration to 70 degrees looks about right). We can even have the robot wait for the human to acknowledge that the round is over with another push on the Touch Sensor, and then we can use another Motor block for the same port (wired in, again) to rotate the arm in the other direction by the same amount, which would reset the arm. At this point, the sequence would loop back to the beginning, waiting for another press on the Touch Sensor to start the process all over again.

We still need to address the guts of step 2. Here we can have a little bit more fun. On each throw, we'll have RaSPy display the number on the LCD, as well as say the number out loud, while at the same time moving all three arms down and back up. First we need to figure out which throw we're on—the first, second, or third. We can get that information from the counter plug that we exposed on the inner loop as shown in Figure 9-4, but (as Jim once

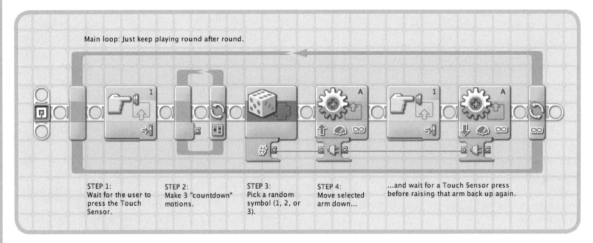

Main loop: Just keep playing round after round.

STEP 1:
Wait for the user to press the Touch Sensor.

STEP 2:
Make 3 "countdown" motions.

STEP 3:
Pick a random symbol (1, 2, or 3).

STEP 4:
Move selected arm down...

...and wait for a Touch Sensor press before raising that arm back up again.

Figure 9-3: Give RaSPy the ability to mimic the game's opening.

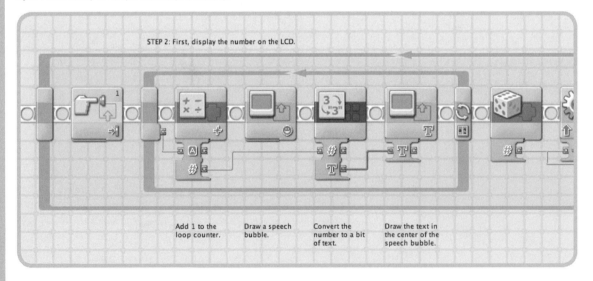

STEP 2: First, display the number on the LCD.

Add 1 to the loop counter.

Draw a speech bubble.

Convert the number to a bit of text.

Draw the text in the center of the speech bubble.

Figure 9-4: We also display the count on the LCD.

informed me) computers start counting with zero, so in order to get RaSPy to display *1* on the first iteration through the loop, we need to add one to whatever is wired out of the loop counter. Next we can display a speech bubble graphic on the LCD, and (*without* clearing the LCD again) display the number (suitably changed into a bit of text for the Display block). Figures 9-5 and 9-6 show how to configure the Display block to show both a speech bubble and the count centered in that bubble.

Getting RaSPy to speak the number is really easy—all we have to do is wire the number into a Switch structure (one configured to switch on a value, specifically on a number) and put a different Sound block set to say a different number in each case of the switch. This is demonstrated in Figure 9-7. (These Sound blocks are conveniently provided by LEGO right in the software.) This means we have to configure a multicase Switch structure—if you need help with that, take a look at "Loop and Switch Structures" on page 13. Also, if you

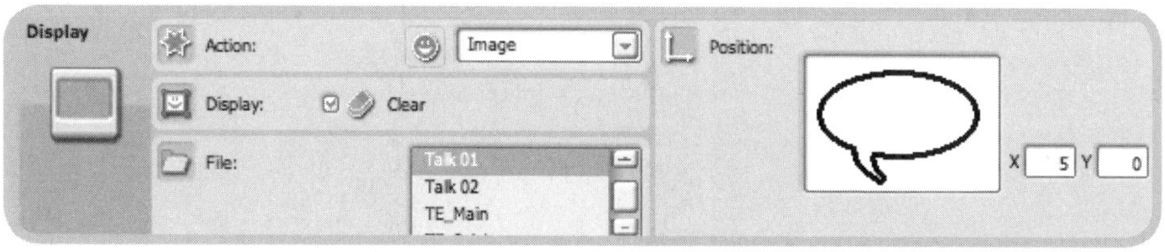

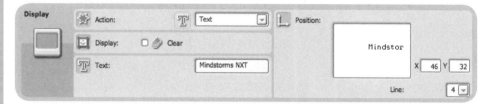

Figure 9-5: RaSPy's count will be displayed in a speech bubble.

Figure 9-6: Centering the count inside the speech bubble

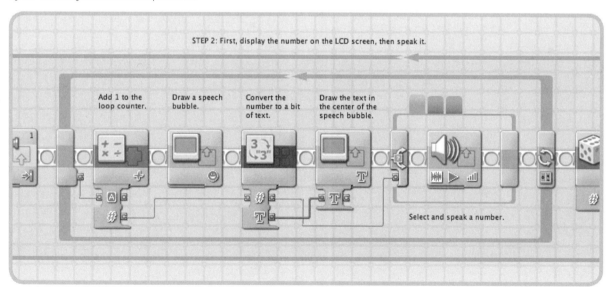

Figure 9-7: The Switch and Sound blocks will allow RaSPy to count out loud.

set the Sound blocks to not wait for completion, then the spoken phrases will occur at the same time as the other actions, like the arm movements. Figures 9-8 and 9-9 show the proper configuration of the Switch and Sound blocks to get RaSPy to "talk."

All that's left is to make all three arms go up and down together. This is a little bit tricky; the proper placements of the Motor blocks are shown in Figures 9-10 through 9-14. For one thing, even the Move block only synchronizes two motors, not all three, and furthermore, each arm is different, so each will move at a slightly different speed at a given power level. One way we can synchronize the arms is by using three parallel sequences—one moving motor A back and forth, a separate sequence for motor B, and a third for motor C. With all three of these running in parallel, the motions should take place at roughly the same time.

To account for the different weight of each arm, we can set the Motor blocks to try to control the power used: The NXT will try to adjust the power up or down to make for a motion that looks like it is at a constant unloaded power level. To put it another way, it functions a little bit like a speed control, so that the speed of a motor set to, say, a power of 50, will always be about the same, regardless of whether the motor is unloaded or moving a significant load. The other tricky part about this, of course, is making room for all three sequences in that little itty-bitty loop (apply the crowbar trick from "When to Use a Crowbar in Programming" on page 24, shown in Figures 9-10 through 9-13). Once you have the loop opened wide enough, just lay down some Motor blocks in pairs: one to move an arm 45 degrees one way, and a second to move it 45 degrees back.

Figure 9-8: The Switch block configuration panel has three options.

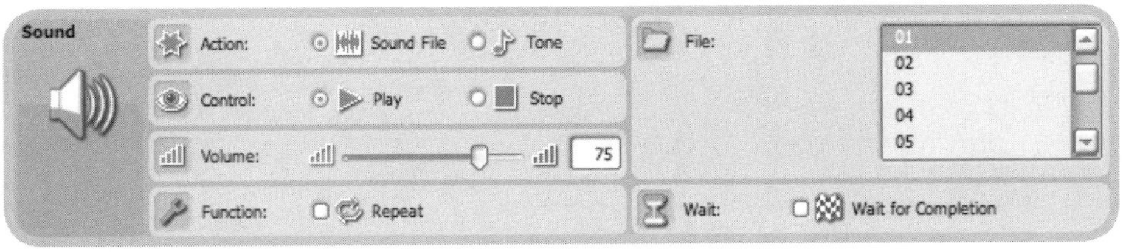

Figure 9-9: Use the Sound block configuration panel to pick 1, 2, and 3 for speech.

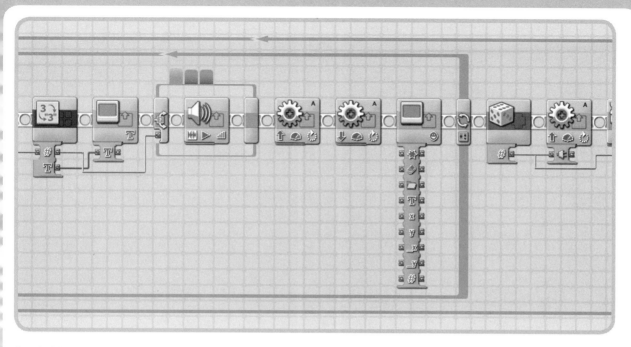

Figure 9-10: Dropping the first two Motor blocks and a Display block as a crowbar

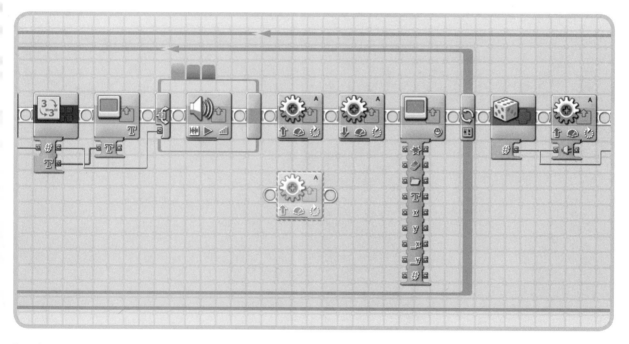

Figure 9-11: Dropping an orphan Motor block to start the second sequence

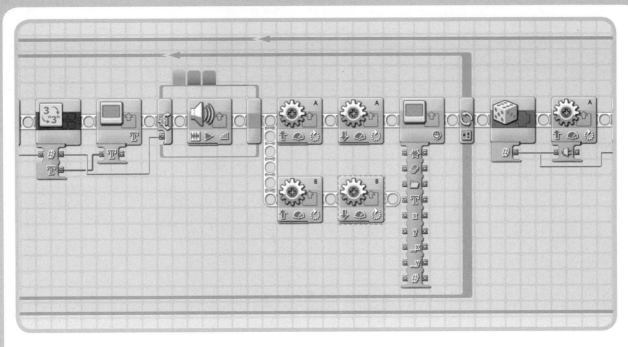

Figure 9-12: Completing the second sequence by SHIFT-clicking to drag out a branch of the sequence beam

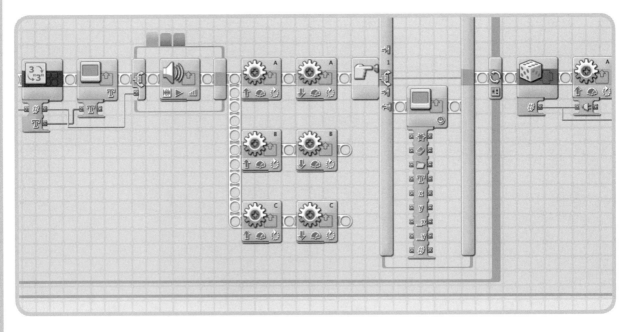

Figure 9-13: Placing the third sequence using a nested crowbar to open the loop wider

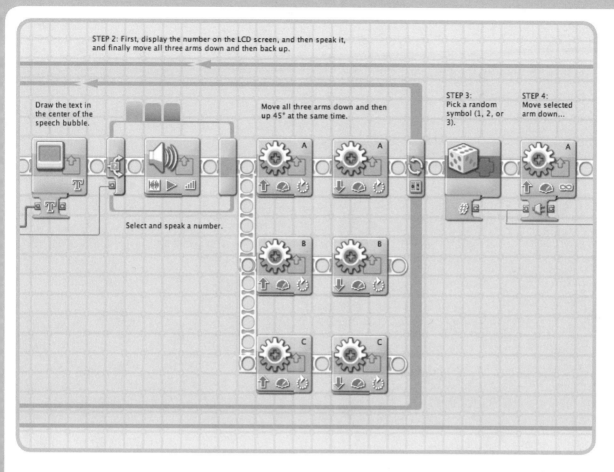

STEP 2: First, display the number on the LCD screen, and then speak it, and finally move all three arms down and then back up.

Draw the text in the center of the speech bubble.

Move all three arms down and then up 45° at the same time.

STEP 3: Pick a random symbol (1, 2, or 3).

STEP 4: Move selected arm down...

Select and speak a number.

Figure 9-14: Finished sequences with the crowbar removed

That's it. The finished program is shown in Figure 9-14. Make sure you have all the sensors and motors hooked up to the correct ports and all the ports specified correctly in your program, and you should be good to go! You can play Rock, Scissors, Paper to your heart's content.

Incidentally, there's a reason I left the construction of the parallel sequences until the very end. While NXT-G does a good job of reorganizing wires when you add or delete blocks, it's not nearly as graceful with sequence beams (or comments, for that matter). As an example, once the program is complete, try deleting the Number to Text block, and you'll see that the parallel sequences (including the point at which they appear to branch) end up being scrambled quite a bit. A similar thing happens if you try to insert a block ahead of those branching sequences, often with unpredictable and undesirable results like blocks or beams getting hidden behind switches or loops. To avoid this, try to make parallel sequences some of the last things you do in a program. Likewise, while commenting is a very good thing, keep in mind that comments are not tied to any particular block—so as a program grows or shrinks on the screen, the comments may not end up anywhere near the specific block you originally placed them with. Save detailed comments

and branching sequence beams for near the end of constructing a program, if possible.

Is that it for RaSPy? Well, that depends on you. While this is a simple example, there are a number of places you could take it. For instance, right now RaSPy has to trust you completely: Not only does he not keep score, but he doesn't even know if you are actually playing, and he certainly has no way to judge who wins. If instead of just throwing out your hand in some form that RaSPy can't "see" anyway, if you pushed one of the three front-panel buttons at the right time, RaSPy could have feedback, knowing which choice you made—and with a little more programming, he could even detect if you tried to cheat by waiting a little too long and seeing what he chose. Once RaSPy had such feedback, he could keep score as well, keeping a running tally of who is ahead and by how much. I can even think of other ways to play the game—for instance, have the NXT display its choice as a symbol on the LCD, while you have to submit your choice by pushing a joystick. (Left for rock, middle for scissors, and right for paper, perhaps?) Don't let this little experiment end here. We've given you the model, now have fun making it better.

beach buggy chair:
a ramblin' robot

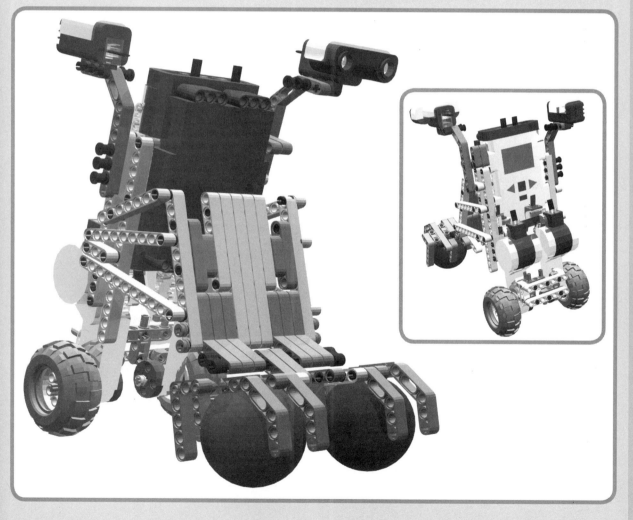

Figure 10-1: The beach buggy chair (front and rear views)

Since robots are becoming an increasingly important aid to the disabled, we decided to construct a beach buggy chair with our basic MINDSTORMS NXT kit. *Beach buggy chairs* are real motorized chairs that allow people with disabilities to navigate ocean beaches. If you spend time at the beach, you may have seen a beach buggy chair in action.

The two balls in the front of our buggy give us the sense of balloon tires, and it has enough of a chair to give a small teddy bear a wild ride around the house. (Your passenger might need a seat belt to stay on, though.)

NOTE Do not use this buggy on a real beach. The sand and moisture would damage your motors and the NXT brick.

building the beach buggy chair

The Beach Buggy Chair (Figure 10-1) is a rather simple robot, but here are some things to keep in mind:

* When you connect the wheel hubs to the seat back, you will be inserting the longest axle in your kit through some tight spaces. It will require some serious pressure and care. (Hint: When you disassemble the Buggy, use another long axle to push the axle out.)
* It's important to make sure the Ultrasonic Sensor is pointing forward and slightly down. If you follow these instructions exactly, it should end up in the correct position.
* Even though the wheel in the illustration looks slightly different from the one in your kit, the wheel in your kit is the correct one.

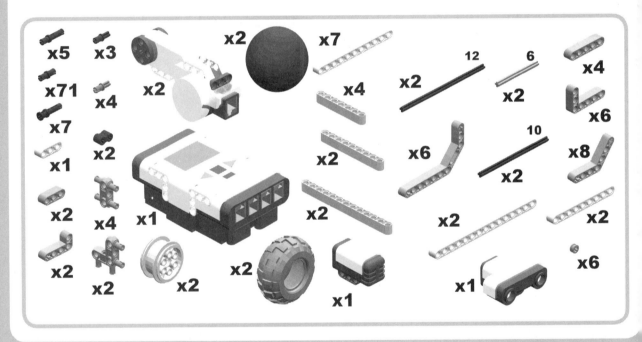

seat back

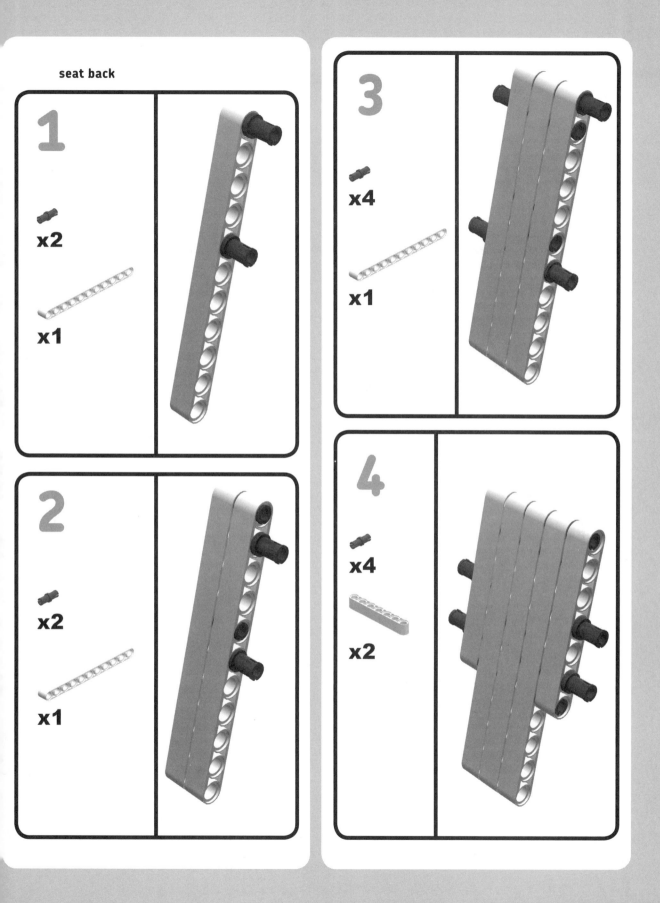

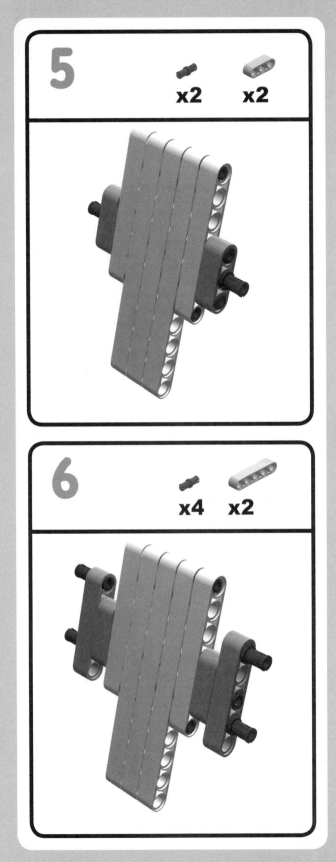

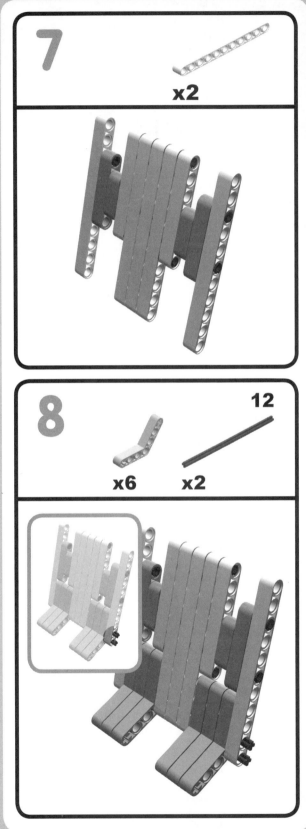

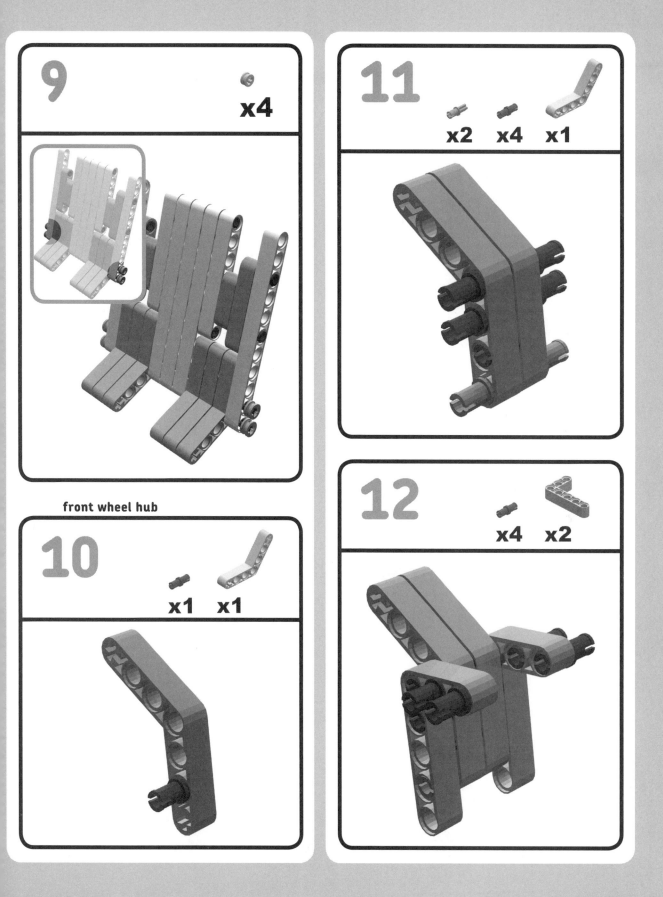

9 x4

front wheel hub

10 x1 x1

11 x2 x4 x1

12 x4 x2

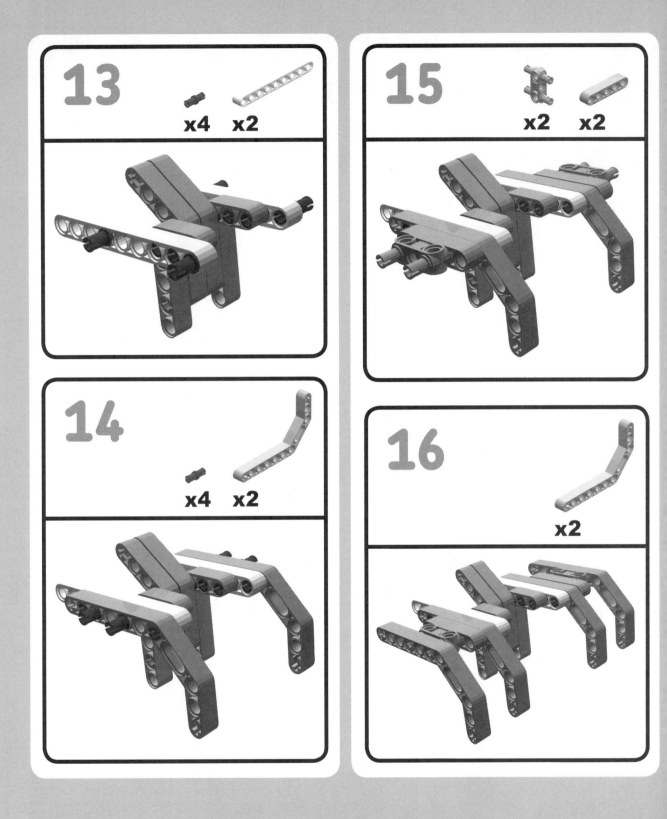

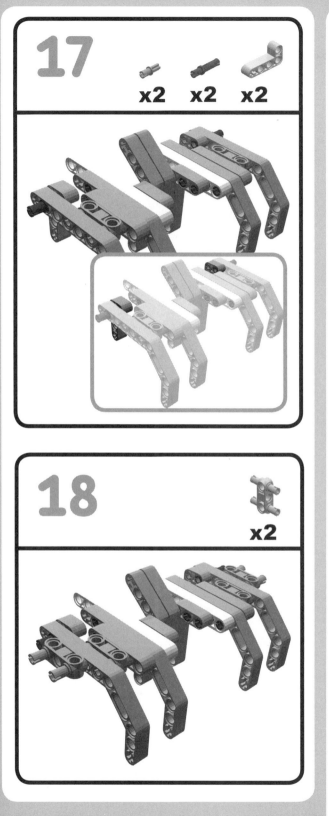

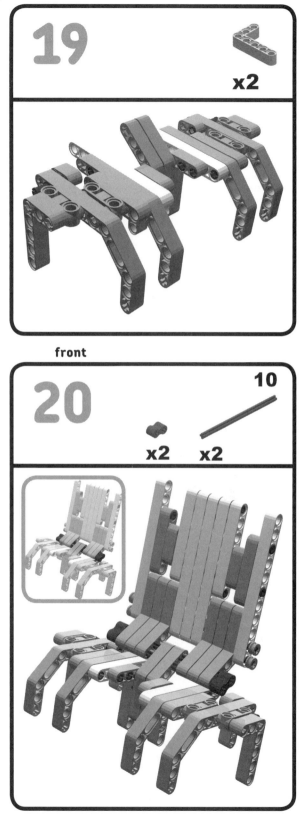

front

robot back

21 x2

22 6

x2 x2 x2

23 x2

24 x8

x1 x1

x1

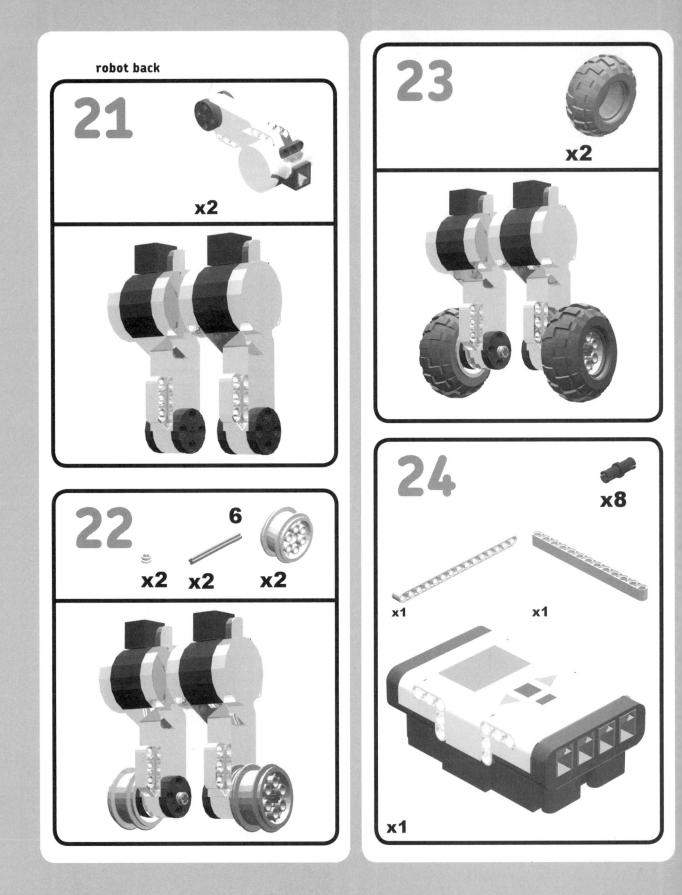

24

25

x8

x1   x1

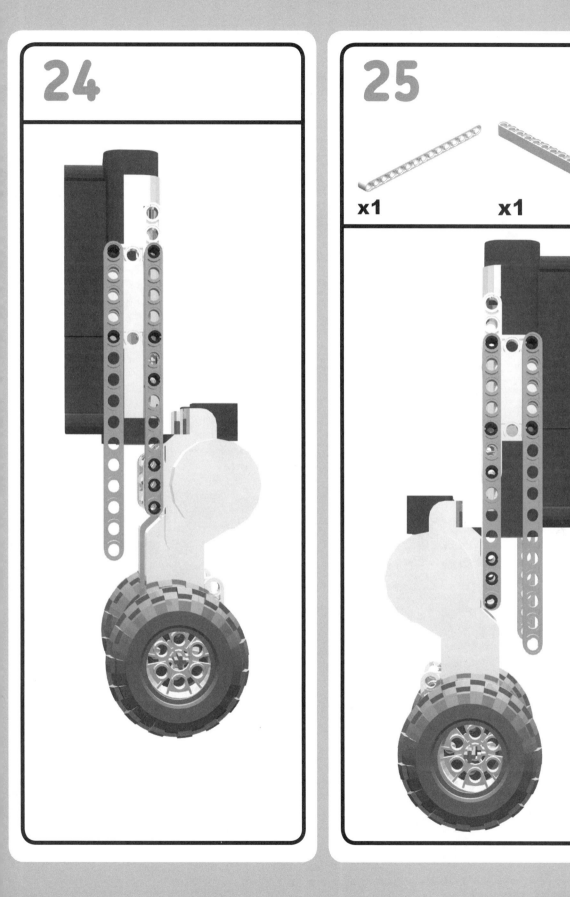

26

x2

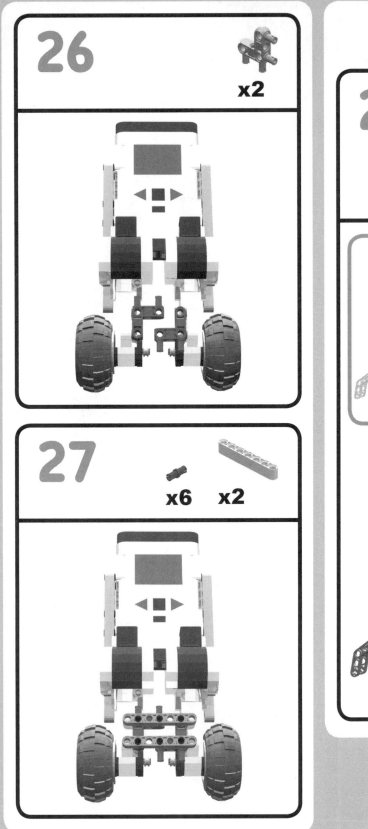

27

x6 x2

**final assembly
(including sensors)**

28

x1 x8

x1 x1

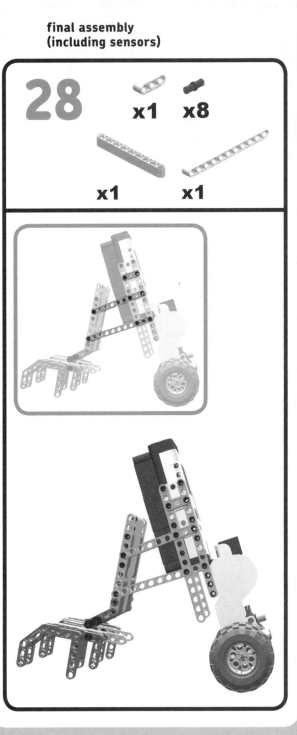

29

x2   x2

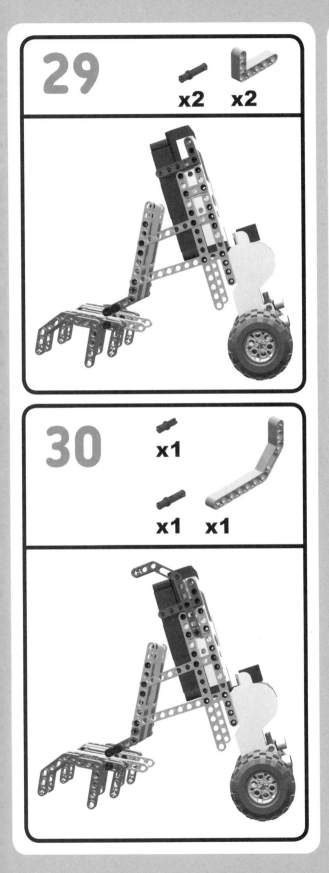

30

x1

x1   x1

31

x1

x2   x1

32

x6 x1 x1

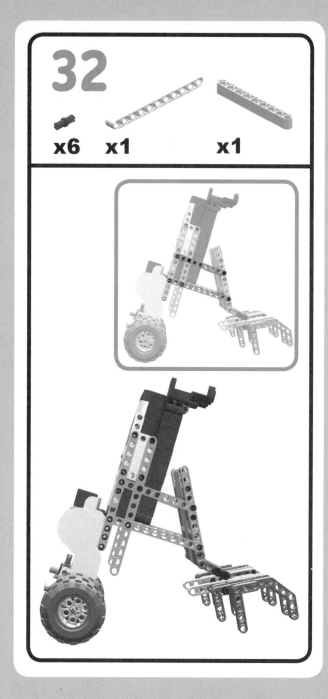

33

x3 x1

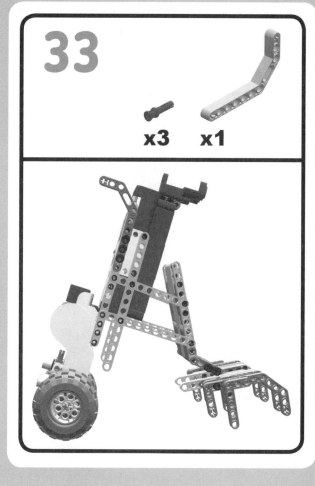

34

x1 x2 x1

(Sound Sensor)

power connections

This model requires four power cables.

The Ultrasonic Sensor is connected to input port 4 at the bottom of the NXT brick. The Sound Sensor is plugged into input port 2. The cables in the motors are connected to output ports B and C on top of the brick.

35

x6

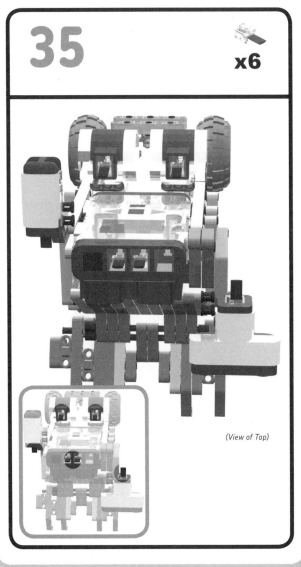

(View of Top)

36

x2

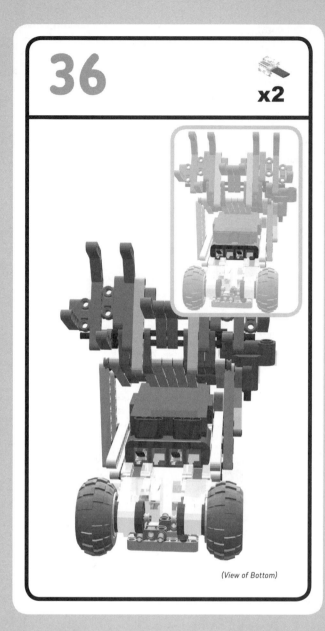

(View of Bottom)

37

B   **R**

x1   x1

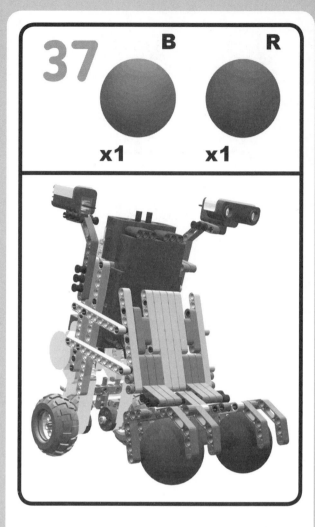

There is one red ball and one blue ball in your NXT kit. Your beach buggy chair is now complete.

programming the beach buggy chair

Our first goal in writing this program was to prevent the buggy from getting stuck in corners. (This is a notorious problem in most programs we've tried.)

Our second goal was to be able to start and stop the vehicle with sound—something a disabled person might need to do.

my blocks

The first thing we want to do is to create some custom My Blocks. *My Blocks* are your own building blocks for NXT programming. With My Blocks, you create small bits of programming that can be used repeatedly in any other program you might write. In a way, they allow you to program in shorthand so you can view a complicated program in simpler form on your screen. (For an introduction to My Blocks, see "My Blocks Save Time and Simplify Your Programs" on page 16.)

Without My Blocks, your program would become an ever-expanding mass of switches and loops, branching out of sight. If you don't use My Blocks, you'll likely be consigned to constantly dragging the screen around in order to see and modify its parts, and it will be very difficult to keep track of everything.

NOTE The inside cover of this book contains full-color illustrations of all available programming blocks.

my block #1—motormove60

This block tells motors B and C, individually, to move forward. *Remember, in this robot, the motors are positioned in such a way that you must change the default direction setting if you want them to move the buggy forward.* Because we want the motors to turn at the same time, we'll put Motor blocks on parallel beams.

1. Open a new document (Figure 10-2).

2. Select a Motor block and drag it to the Start (Figure 10-3). In the configuration panel at the bottom of the screen, select port B and a power setting of 60. Leave the default settings for everything else.

3. Place a second Motor block a few spaces below the first block (Figure 10-4).

Figure 10-2: The new document

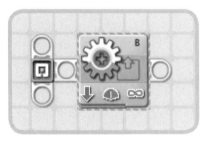

Figure 10-3: The Motor block

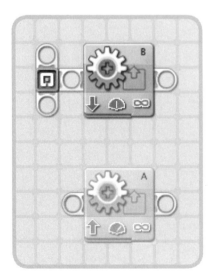

Figure 10-4: The second Motor block

4. Hold down the SHIFT key and use your mouse to drag a beam to the second Motor block (Figure 10-5). (This may take some practice.)

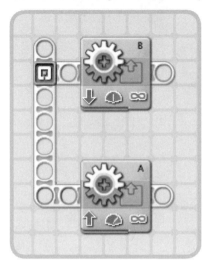

Figure 10-5: Dragging the beam

5. Select the second block and configure it for port C, with power set to 60 (Figure 10-6). *It's important that both motors have the same power setting.*

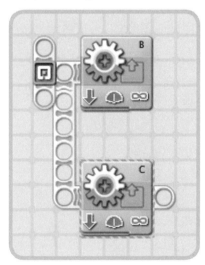

Figure 10-6: Configuring the second block

6. Now we'll save this as a My Block. Use your mouse to select all the blocks and beams, and click the **Create My Block** symbol at the top of the screen (Figure 10-7).

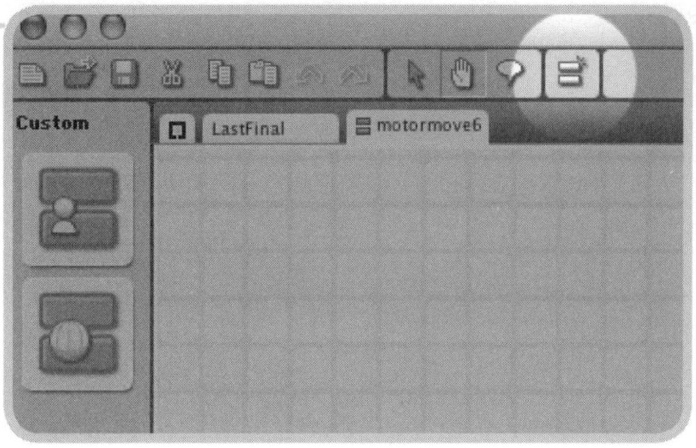

Figure 10-7: The Create My Block button

7. A dialog box will appear asking you to name your block and giving you an opportunity to write a description of it (Figure 10-8). Name this block *motormove60*.

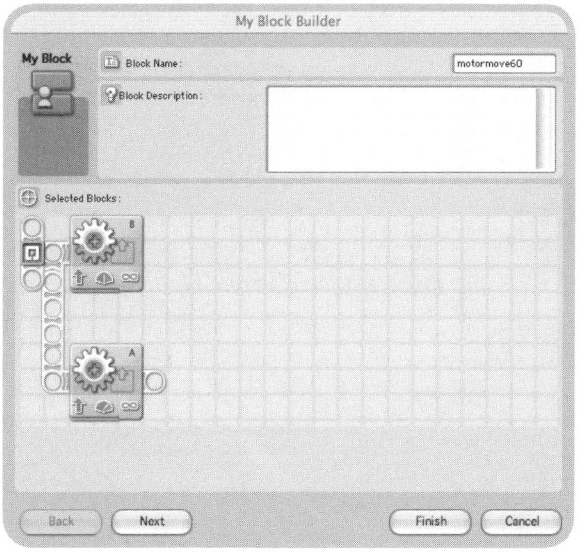

Figure 10-8: Naming the My Block

8. Click **Next**. This dialog box allows you to associate an icon with your block to help you recognize it among all the other My Blocks (Figure 10-9). We've chosen a gear; you can pick anything you like for yours.

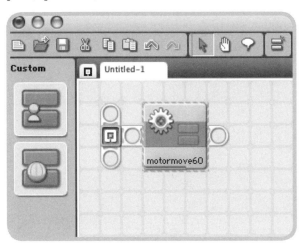

Figure 10-9: Choosing an icon for the My Block

9. Click **Finish**. Your My Block is now available for use in any program (Figure 10-10).

Figure 10-10: The motormove60 My Block

my block #2—tightrotate

Turning space is a factor when a robot ends up in a corner. There is little room to maneuver when a robot gets itself into a corner. To give our buggy the best chance of getting out of a tight spot, we'll program both motors (B and C) to move the same number of degrees in opposite directions. This causes the robot to turn in a tighter loop than if we'd just engaged one motor.

1. Open a new document.

2. Select a Motor block and drag it to the Start (Figure 10-11).

Figure 10-11: The Motor block

3. Place a second Motor block a few spaces below the first block (Figure 10-12).

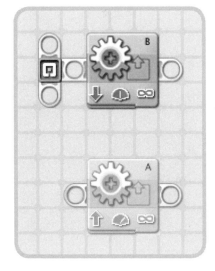

Figure 10-12: The second Motor block

4. Hold down the SHIFT key and use your mouse to drag a beam down from the first Motor block to the second one.

5. Select the top block (Figure 10-13). Configure it to port B with the power set to 60. Next to the word *Duration*, select **Degrees** from the drop-down menu and type 90 as the amount. (This won't actually turn the robot 90 degrees, but it serves our purpose.) We'll accept the default direction for this block.

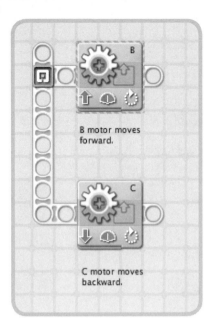

Figure 10-13: Dragging the beam

6. Select the bottom Motor block (Figure 10-14). Configure it to port C with the power set to 60. Change the direction by selecting the down arrow next to the word *Direction*. Choose **Degrees** again from the Duration drop-down menu and type 90 as the amount.

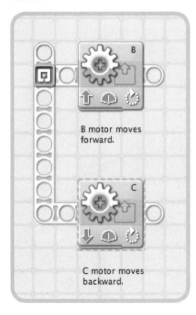

Figure 10-14: Configuring the blocks

7. With your mouse, select all the program blocks and click the **Create My Block** symbol at the top of the screen.

8. Name this block *tightrotate*.

my block #3—55chkloop

This loop uses the Ultrasonic Sensor to find a direction that has at least 55 centimeters of open space.

1. Open a new document.

2. Select a Loop block and drag it to the Start.

3. Configure it as an Ultrasonic Sensor with a distance of >55 cm (Figure 10-15). (Make sure you select **Centimeters** in the drop-down menu next to the word *Show*.)

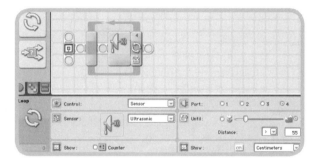

Figure 10-15: Configuring the Loop block for the Ultrasonic Sensor

4. Select the My Block palette by clicking the turquoise parallel bars near the lower-left side of your screen. When this palette opens, you'll see two turquoise blocks. Click the top block to see your My Block library (Figure 10-16). (The name of each block appears when you mouse over it.)

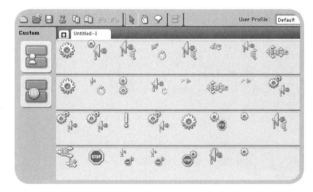

Figure 10-16: The My Block library

5. Select **tightrotate** from the My Block palette and place it inside the Ultrasonic Sensor loop (Figure 10-17).

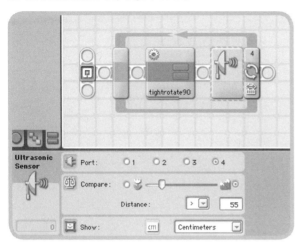

Figure 10-17: The tightrotate My Block

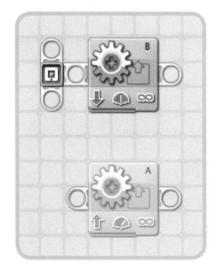

Figure 10-19: The second Motor block

6. With your mouse, select all the program blocks and click the **Create My Block** symbol at the top of the screen.

7. Name this block *55chkloop*.

my block #4—motorstop

This block will stop the motors.

1. Open a new document.

2. Select a Motor block and drag it to the Start (Figure 10-18).

Figure 10-18: The Motor block

3. Place a second Motor block a few spaces below the first block (Figure 10-19).

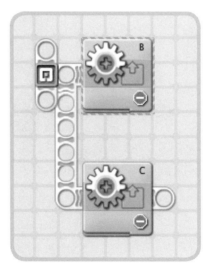

Figure 10-20: Dragging the beam

4. Hold down the SHIFT key and use your mouse to drag a beam down from the NXT square to the bottom block (Figure 10-20).

5. Configure one block as port B and the other as port C. Choose the stop symbol next to the word *Direction* for both blocks.

6. Select all the instructions and click the **Create My Block** symbol.

7. Name this block *motorstop*.

putting the pieces together

Now we're ready to construct our program. Since we want to start action with a sound, the first command is to listen for a loud sound. To do this:

1. First place a Wait block on the beam. (The Wait block is the block with the hourglass symbol on its face, as shown in Figure 10-21.)

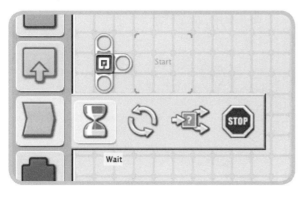

Figure 10-21: The Wait block

2. Configure the Wait block as a Sound Sensor with level >50 (Figure 10-22).

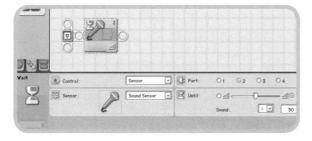

Figure 10-22: Configuring the Wait block

3. Place motormove60 from your My Blocks on the beam next to the Wait/Sound block (Figure 10-23).

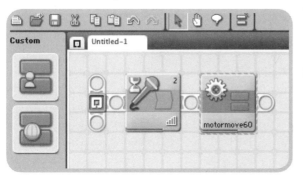

Figure 10-23: Placing the My Block

4. Next we'll add a Loop block. Configure the loop as a Sound Sensor with a level of >85 (Figure 10-24). This setting is one you'll need to personalize. At the default setting of >50, our robot turned itself off with its own noise. However, setting the level too high made it difficult to stop the buggy. Just be aware that you may need to tinker with this setting before your robot cooperates.

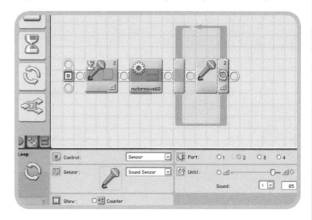

Figure 10-24: Configuring the Loop block

5. Inside the loop, place a Wait block configured as an Ultrasonic Sensor with a distance of <45 centimeters (Figure 10-25).

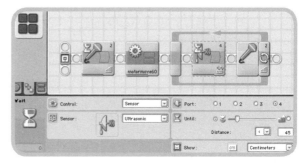

Figure 10-25: Configuring another Loop block

6. Follow that with My Blocks 55chkloop and motormove60 (still within the loop) and then add a motorstop My Block outside the loop (Figure 10-26).

   Now you're ready to download the program and take your buggy for a test drive.

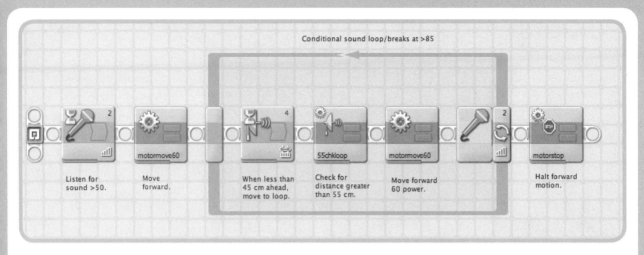

Figure 10-26: Complete program with My Blocks

troubleshooting tips

* We expected that the sound loop would simply turn off the robot when it heard a loud sound. In fact, the Sound Sensor does not respond to a sound cue at every point in the NXT sound loop. In order to end our buggy program, we often had to extend or repeat our sound cue ("Stop! Stop! Stop! Stop! Stop! Stop! Stop!"). Also, when the Sound Sensor is set at a high level (as ours needed to be), we had to get close to the sensor to stop the robot.

* The Ultrasonic Sensor also has limitations. For example, it can be fooled by some surfaces it encounters. While testing this program, we discovered that our sensor sometimes did not recognize a solid wall when approaching it from an angle. We assume this has something to do with the reflection of the sound waves on that particular surface. Sometimes you can improve accuracy by changing the angle of your sensor, but it will inevitably fail periodically.

* Setting the sound and power levels for your blocks is not an exact science. What works well one day might not work well the next. What works on hardwood floors may not work on carpet. For months, our robot responded very well to sound cues of >50. Suddenly, one day it started turning itself off and we had to increase the sound levels dramatically—same room, same buggy, different batteries. Did using lithium batteries make our buggy run louder, as well as stronger? Whatever the reason, the settings that had worked well for so long no longer worked.

* Beware of setting your power levels too high. Even though a fast robot may look cool, it's not so cool to see it run into walls because a sensor didn't have time to respond and stop it. But when we run our buggy on the carpet, the power settings must be increased.

* For a long time, we misunderstood the rotation degree settings for turning the motors. We thought that setting the motors to turn 90 degrees meant they would make 90-degree turns. Actually, this setting only makes them turn something like 45 degrees. Consider this when you design your programs. (For more information on how to make turns, see the instructions for Marty in Chapter 16.)

hints for further exploration

If you have no interest in warriors, battles, racing, or competition, in general, consider projects with creative or altruistic themes. Robots can dance, sing, and use tools. They can become characters in original stories (as in Jim Kelly's book, *LEGO MINDSTORMS NXT: The Mayan Adventure*, Apress, 2006) or the incarnation of characters from the classics. We recently created a funny video using a robot, that is getting a lot of views on YouTube. It's great fun!

If you're looking for other vehicles for the disabled that you can emulate, check out WheelchairNet (http://www.wheelchairnet.org), an excellent online resource offered by the Department of Rehabilitation Science and Technology at the University of Pittsburgh.

a final thought

Learning robotics is good. Doing good with robotics is even better!

3D PhotoBot:
a 3D photo assistant robot

What in the world is a 3D PhotoBot? Well, most robots have wheels and roll, walk, or crawl. Not this one. This robot has only one purpose, and that is to help you create some excellent three-dimensional photographs using a standard digital camera. Figure 11-1 shows the completed 3D PhotoBot.

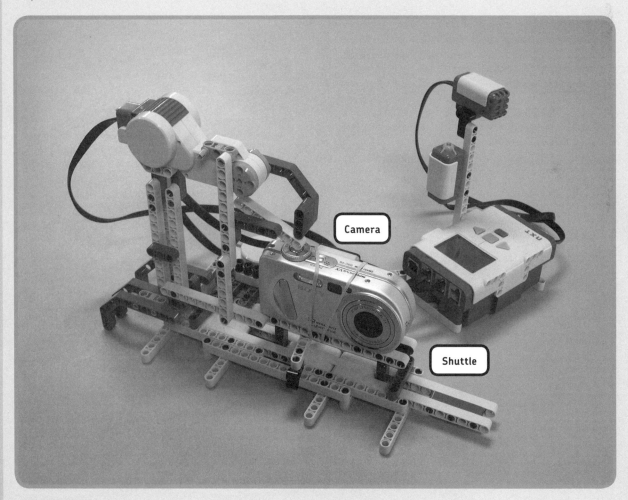

Figure 11-1: Your very own 3D PhotoBot

Notice that the robot has a digital camera sitting on a small, bench-like table that is mounted to a sliding mechanism we call a *shuttle*. The shuttle moves the camera and allows it to take two photographs of the same view.

Our robot uses two NXT motors. One motor moves the camera back and forth, and the other presses the camera's shutter button. We use the Touch Sensor to start the program, and we've added the Sound Sensor so you can make your 3D PhotoBot voice activated!

the art of three- dimensional photography

Once you've built your robot by following the steps in this chapter, you'll place it in a location to take a picture. Ideally, the best three-dimensional images feature something that's close to the camera and something that's in the background. This is the secret to taking three-dimensional pictures. The idea is to build a robot that mimics the way human eyes view surroundings. While human eyes can see three dimensionally at any distance, using pictures to mimic three dimensions has its best results for pictures that contain either close or far away objects. This is partly a limitation due to the fact that you are taking a two-dimensional medium (photos, in this case) and attempting to force your human eyes to see a three-dimensional image. We suggest you try a variety of tests and distances to determine the best results for your robot.

In order to create three-dimensional photos, you need to take two images of the same object, from two different angles.

The average set of human eyes has about 2.5 inches (6.35 cm) between each eye (measuring from the pupils), which means that each eye looks at the same object from a slightly different position. To get a feel for this, try this experiment:

1. Look at a nearby object with both eyes open.

2. Close your right eye and look at the same object.

3. Next, open your right eye and close your left eye.

You should notice a slight shift in the object's position. Each eye is looking at the same object, but from a slightly different position. When your brain takes these two different images and blends them together, the result is three-dimensional vision (sometimes called *stereo-optic vision*).

The 3D PhotoBot produces three-dimensional photos by taking photographs of objects from two different angles, separated by 2.5 inches—about the average distance between human eyes.

For example, Figure 11-2 shows two images that look almost identical; but if you take a closer look, you'll see some differences. These images were both taken using the 3D PhotoBot. The image on the right was taken using the digital camera in position 1 (consider this to be your right eye). The 3D PhotoBot then moved the camera 2.5 inches (6.35 cm) to the left and took the second photo at position 2 (consider this to be your left eye). Everybody's eyes are different, so if you want to mimic your own eyes, just have a friend measure the distance from your left pupil to your right pupil. The robot will be programmed to move from right to left 2.5 inches (6.35 cm) to simulate this distance and will provide two pictures—a simulated "left eye" picture and a "right eye" picture. These will be used to create a three-dimensional image.

Figure 11-2: Two photos taken 2.5 inches apart

producing three-dimensional images

Once we have our two images, we need to turn them into one three-dimensional image.

First, we need a small, single-sided mirror with slightly larger dimensions than the photos. For example, if you will be taking 3-by-5 inch photographs, try a mirror size of 4 by 6 inches.

NOTE The good news is that a 4-by-6 inch mirror will cost about $1 and can usually be obtained at any glass shop. Call around to a few local glass shops and ask them how much they charge for a small mirror. You might even consider getting a collection of mirrors in different sizes so you can experiment with creating different sized three-dimensional images.

Next, upload your digital images to your computer, and name each according to its angle of attack (for example, picture_right.jpg and picture_left.jpg, or something similar).

Now, here comes the trick. Using a graphics program, horizontally flip the picture taken from the right side to change it into a mirror image of the original. The resulting two images would now look something like Figure 11-3. Don't worry if words in the picture are reversed—we'll fix them later.

Once you've flipped one of your images, combine the two into one image that looks similar to Figure 11-4. To do so, either print each image and combine the prints manually, or open both images in a graphics program and place the left edge of picture_right.jpg against the right edge of picture_left.jpg (as shown in Figure 11-4); save the combined image as combined_3D.jpg, and print it.

Figure 11-3: Two pictures with one (at the right) flipped horizontally

Figure 11-4: Place the two photographs side by side, as shown.

viewing your three-dimensional image

Once you have prints of your two mirror-image photos, you'll use a mirror to view them in three dimensions. With your prints on a flat surface, place the mirror's bottom edge directly between the two images, as shown in Figure 11-5. Hold the mirror perpendicular to the photographs so that the mirrored surface is on the right side.

Figure 11-5: Use the mirror to separate the two images, as shown.

Now, holding the mirror still, touch the tip of your nose to the top edge of the mirror. With your right eye, look into the mirror at the reflected image. With your left eye, look at the left image (photograph). You may have to move the mirror slightly to the left or right to make the two images blend together.

When the two images blend, you should see your picture jump out and become three dimensional!

NOTE If you can't see a three-dimensional image, try moving the mirror toward the right image and then away from the right image while keeping the mirror's bottom edge flat on the surface and between the two images. It may also help to place your nose on the left side or non-reflective side of the mirror and then look at the reflection with your right eye.

The reason this works is that your right and left eyes are seeing both images separately and your brain is trying to force both images to look identical. But each image is slightly different because of the original position from which each photograph was taken. Your brain wants to see only one image, so it forces the two slightly different images to blend, which creates the three-dimensional image. What your brain does in the real world is exactly what it is doing here—it takes two different views (with very subtle differences), one obtained via each eye, and then forces them to blend. If you're asking yourself how this works, you'll be happy to hear that scientists and doctors are still baffled by the brain's ability to do this, so don't feel too badly if it seems a bit confusing. Just be thankful your brain can do this; if it couldn't, you'd have to interpret two different views all day while walking around—one from your left eye and one from your right.

notes on using your 3D PhotoBot

There are just a few items to keep in mind when taking pictures with your 3D PhotoBot.

* The 3D PhotoBot works best when taking pictures of objects that do not move.
* If you want to take pictures of people, then you must tell them to stay absolutely still until the second photo is taken.

building the 3D PhotoBot

Building the 3D PhotoBot is not difficult. The trickiest part is probably ensuring that the small arm that will press your camera's shutter button is properly placed over the button. You may need to experiment a few times to get the exact placement. And when the shuttle moves from the left (after taking the first photo) to the right, you'll want to make certain that the small arm doesn't shift and move off of the shutter button.

Follow the instructions as shown and you'll find that you have three main assemblies: the shuttle, the base with one motor to press the shutter button, and the rear portion with the NXT brick and motor to move the shuttle. Assembling the three components is fairly simple, and if you assemble the 3D PhotoBot in the order shown here, you shouldn't encounter any difficulties.

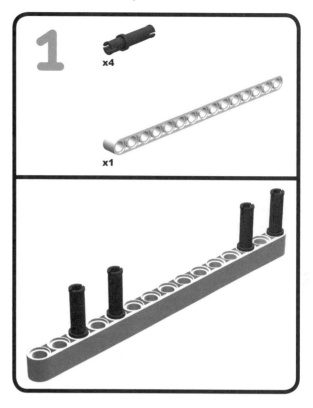

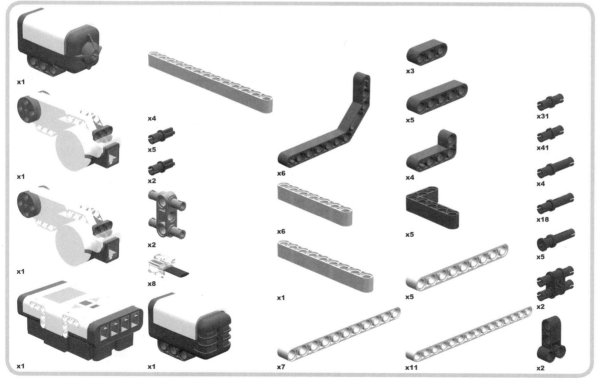

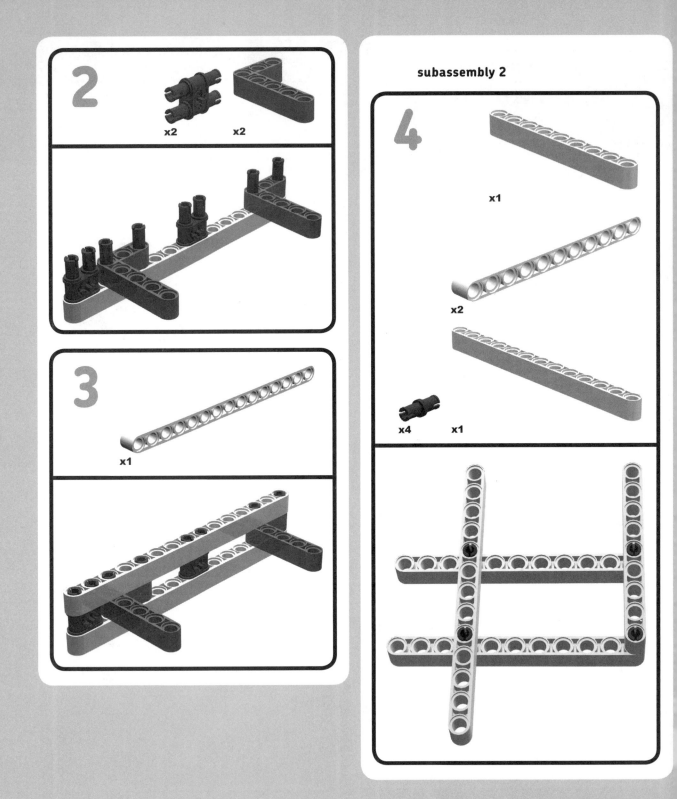

5

x2

x4

x2

6

x6

7

x2

8

x4

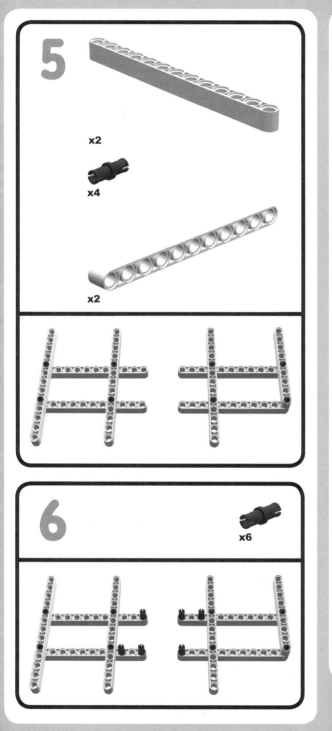

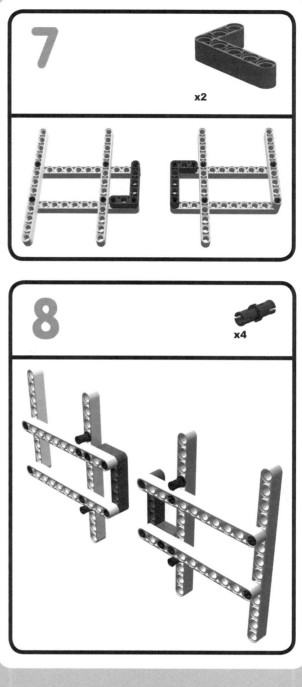

9

x4

x2

10

x2

11

subassembly 3

12

x1

x4

x2

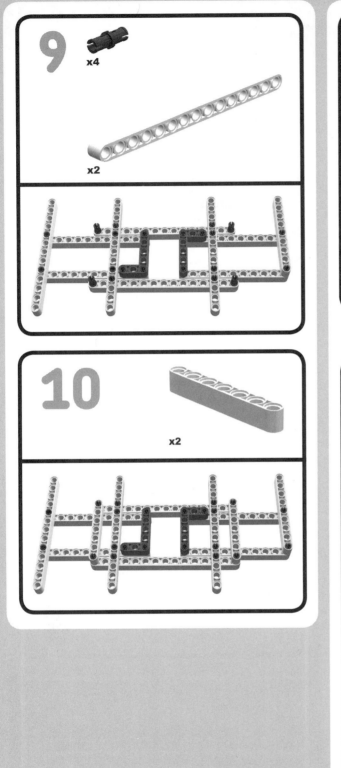
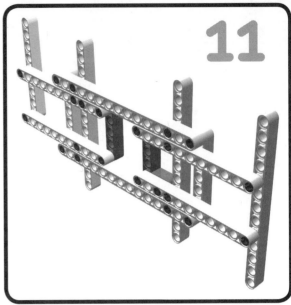
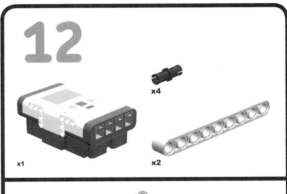

13

x3

14

x4

x1

15

x2   x1

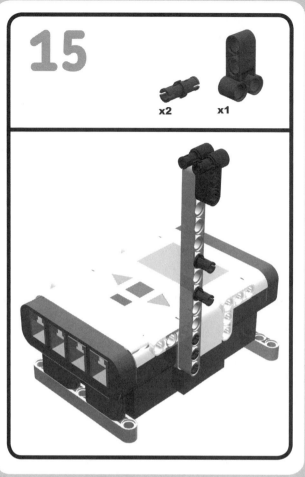

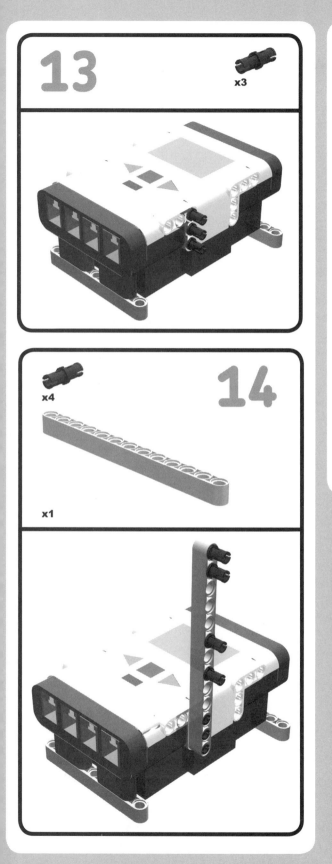

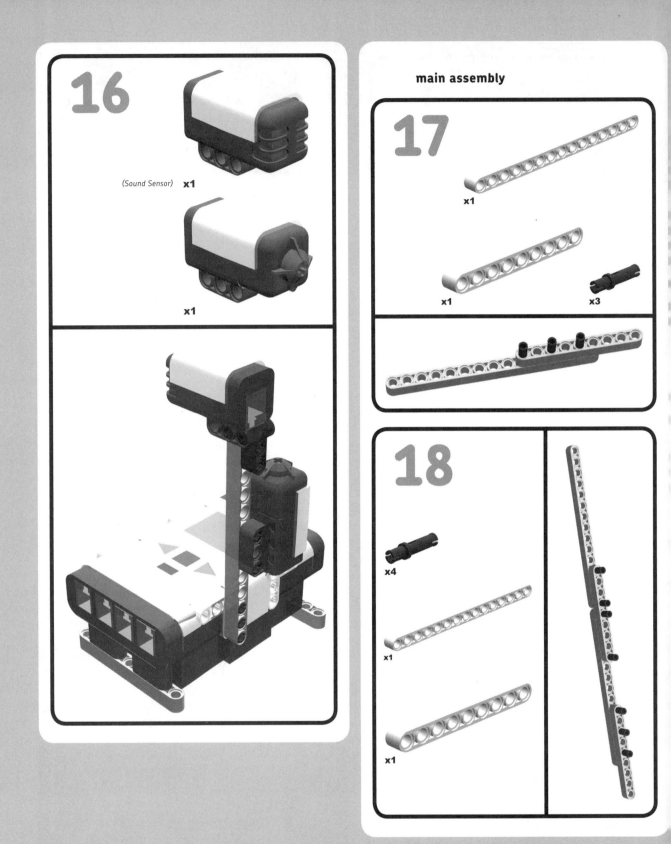

16

(Sound Sensor) **x1**

x1

main assembly

17

x1

x1   **x3**

18

x4

x1

x1

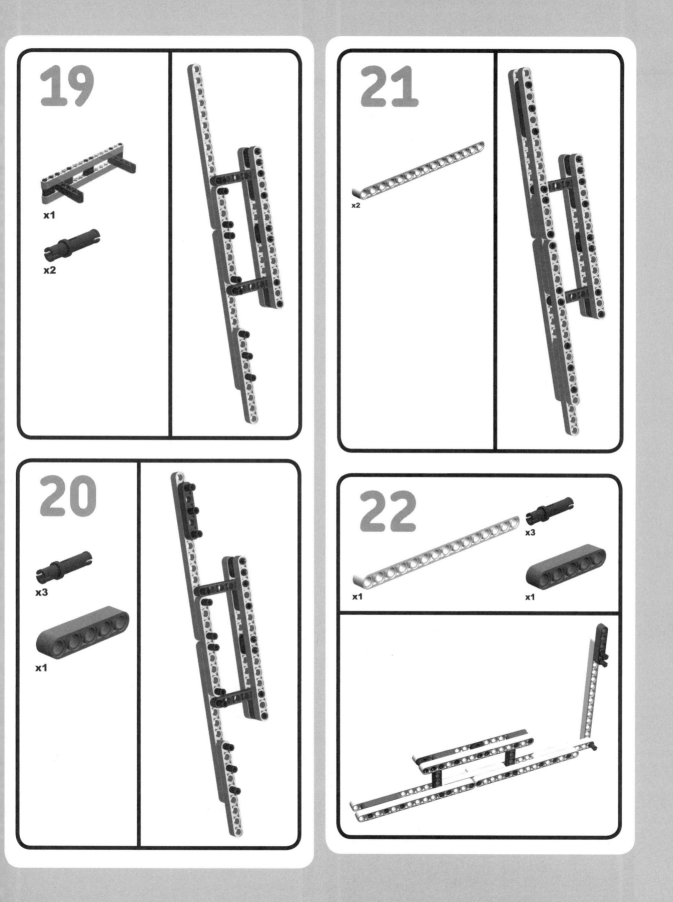

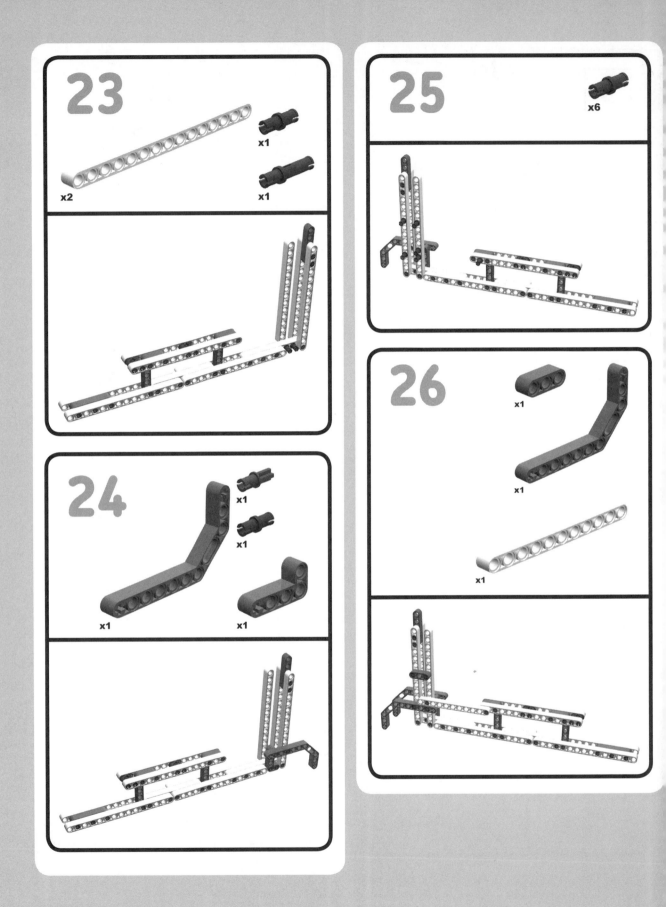

27

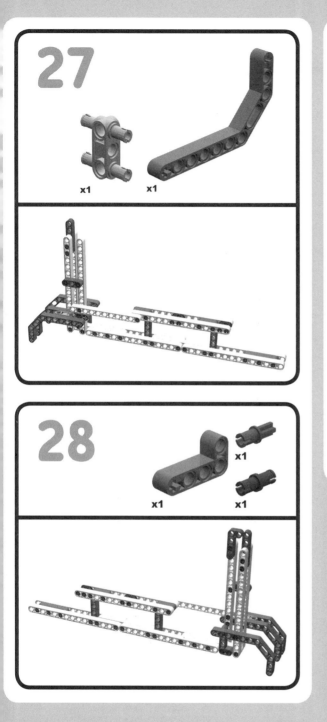

x1   x1

28

x1   x1

x1

29

x2

x1

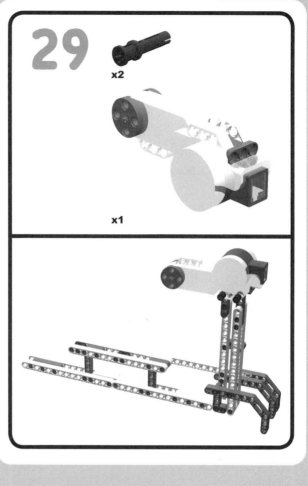

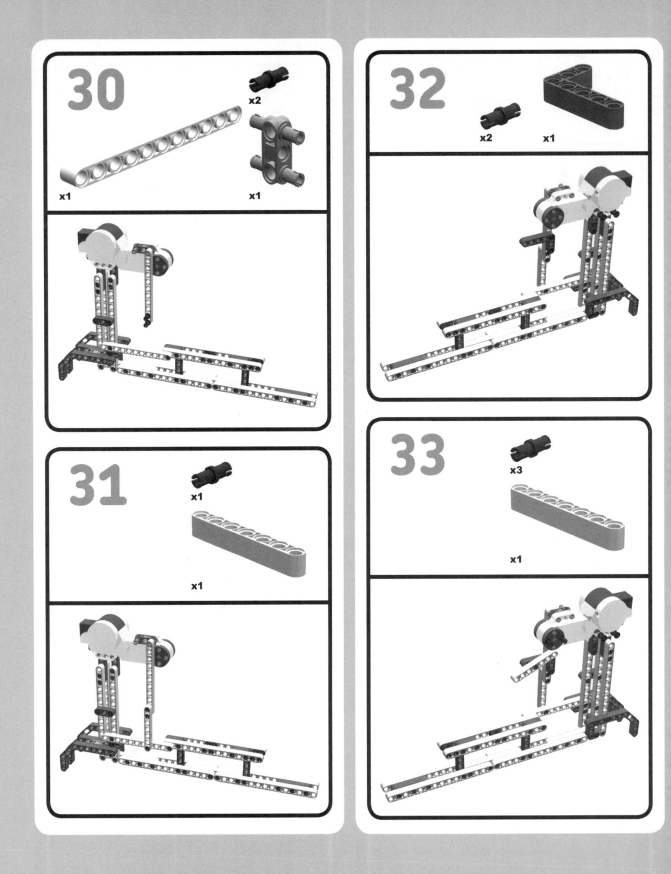

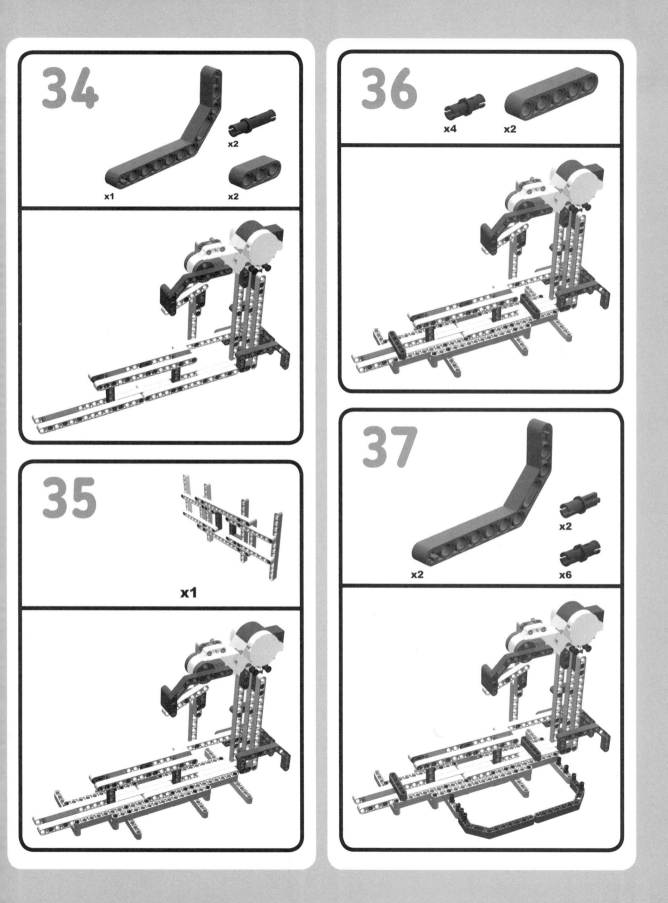

38

x5 x2

x1 x1

39

x1 x2

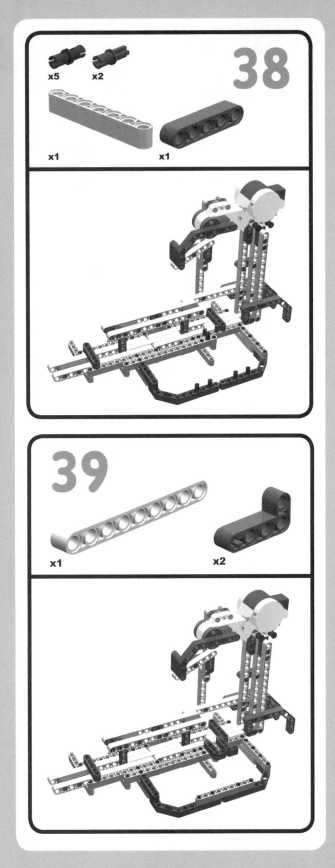

40

x3

x1

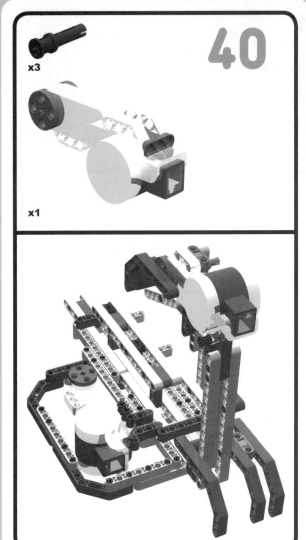

41

x1
x1
x1

42

x2
x1

43

x2
x1
x1

44

x8

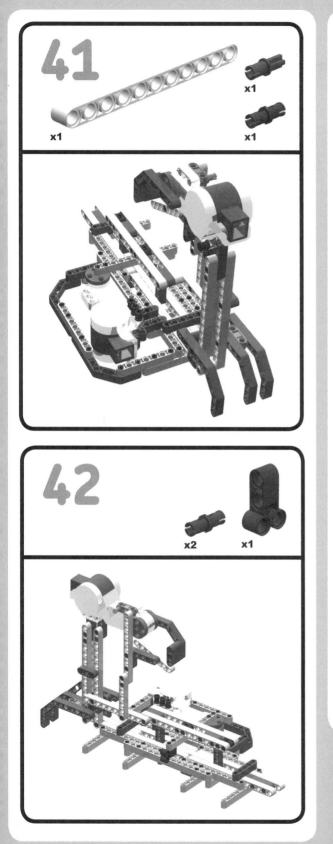

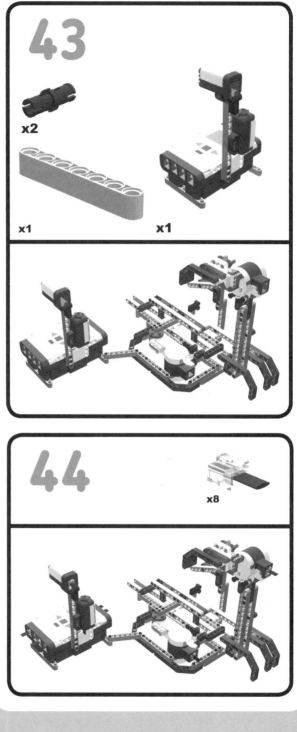

programming the 3D PhotoBot

The program for the 3D PhotoBot uses nine NXT-G blocks. The basic flow of the program follows how the robot actually will perform its task:

1. Move the shuttle's starting position to the far right.

2. After the 3D PhotoBot is turned on, wait for the Touch Sensor to be pressed.

3. Wait for the Sound Sensor to hear *Cheese*.

4. Take the first photograph, and then move the shuttle to the far right.

5. Take the second photograph, and then the program stops.

Below are the steps required to create the basic program.

1. Drag and drop a Wait block configured for the Touch Sensor as shown in Figure 11-6.

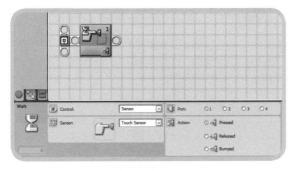

Figure 11-6: The Wait block configured to use the Touch Sensor

NOTE You can eliminate this Wait block if you remove the Touch Sensor from the robot. We've included the Touch Sensor as a second Start button. You begin by running the program (pressing the orange Enter button on the brick) and then orienting the robot to take your desired picture. Once you have the robot placed where you want it, you press the Touch Sensor button to arm the Sound Sensor. If you would rather just run the program and have it immediately become voice activated, you can drop the Touch Sensor from the robot and this initial Wait block from the program.

2. Drop in a Wait block configured to use the Sound Sensor as shown in Figure 11-7.

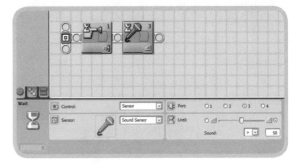

Figure 11-7: The Wait block configured to use the Sound Sensor

This second Wait block will be triggered when the Sound Sensor detects a sufficiently loud sound (such as someone saying "Cheese!"). Experiment with the sensitivity of the Sound Sensor to determine what works best for you.

3. Drop in a Move block configured as shown in Figure 11-8.

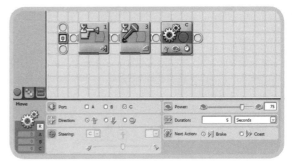

Figure 11-8: The Move block that will take the first picture

This Move block is used to take the first picture. Experiment with the Power setting of the Move block to determine the best speed for pressing the camera's shutter button accurately.

4. Drop in a Move block configured as shown in Figure 11-9.

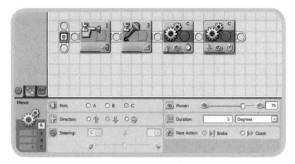

Figure 11-9: The Move block that will release the camera shutter

Once the first picture is taken, you need to release the shutter by moving the small arm a short distance away from the camera's shutter button.

5. Drop in a Wait block configured as shown in Figure 11-10.

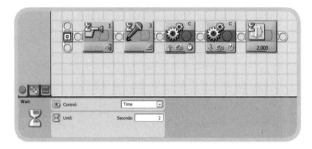

Figure 11-10: The Wait block that prepares the shuttle to move to the left

Once the first picture is taken and the shutter button is released, this Wait block will pause for a few seconds before the shuttle is moved to the left to take the next picture.

6. Drop in a Move block configured as shown in Figure 11-11.

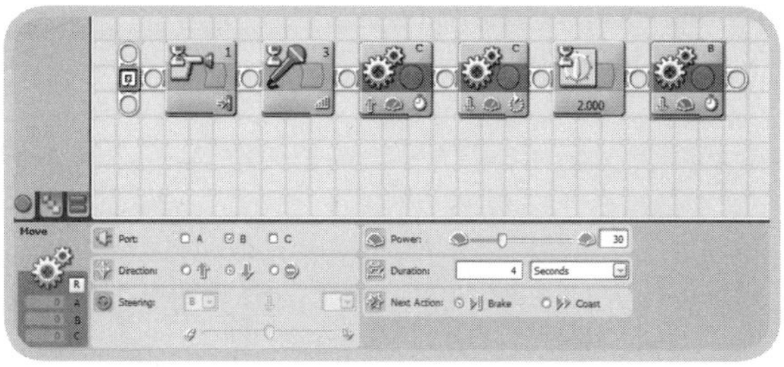

Figure 11-11: The Move block that moves the shuttle to the left

This Move block moves the camera by moving the small swivel arm that pushes the shuttle to the left. Experiment with the settings on this block to find the proper speed and distance to spin the arm that pushes the shuttle.

7. Drop in a Wait block configured as shown in Figure 11-12.

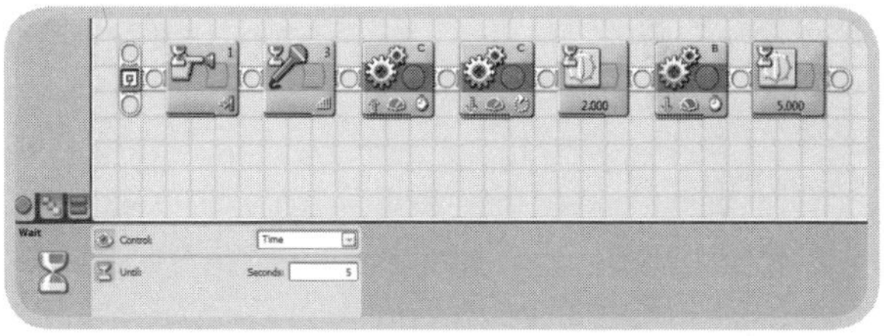

Figure 11-12: The Wait block that prepares the robot to take the second picture

Once the shuttle moves to the left, the robot may wobble a bit due to the shift in weight. This Wait block allows the robot to become completely motionless before taking the next picture.

8. Drop in a Move block configured as shown in Figure 11-13. Now the robot takes the final picture. This Move block is used to press the camera's shutter button. Experiment with the Power setting to determine the value that gives the strongest and most accurate button press.

9. Drop in a Move block configured as shown in Figure 11-14. Finally, the camera's shutter button is released and the robot uses this last Move block to move the small arm away from the shutter button.

After you've created, saved, and uploaded this program, your 3D PhotoBot is ready to take some pictures. Have fun!

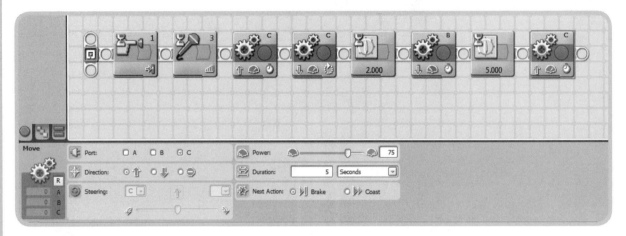

Figure 11-13: The Move block that takes the second picture

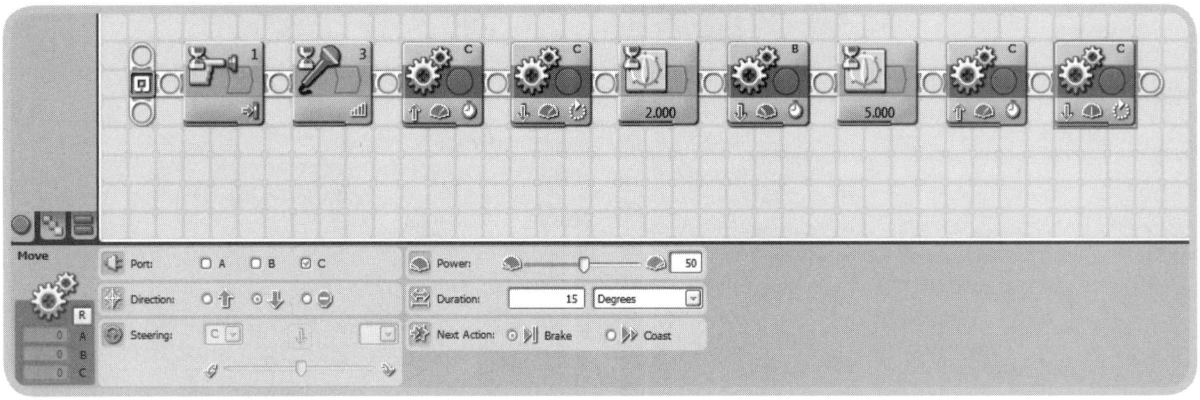

Figure 11-14: The Move block that will release the camera shutter

CraneBot: a grabber robot

One of the first projects many people will tackle with the NXT will be that of building a robot that can move around a room. Having succeeded at that, a good next step is to mount a gripper on the robot so that it can move objects around (Figure 12-1). Grippers appear in many variations because they must be specially designed to match the size and shape of the items they carry.

grippers

The TriBot has a gripper (as shown in Figure 12-2) that is mounted with special parts. As you can see, this gripper seems ideal for grabbing a ball.

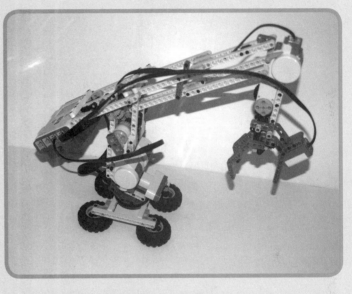

Figure 12-1: CraneBot

Figure 12-2: NXT gripper assembly

However, unlike our hands, while this gripper has a loose grip that works well for gripping balls, it cannot pick up a single LEGO brick and hold it firmly. When we look at our hands, we see the ultimate example of a multipurpose gripper, though with some limitations. Our hands are not really powerful enough on their own to lift heavy or large objects. However, they are capable of very precise movements, which can be very useful for tasks such as writing.

When building a gripper with LEGO, we need to design it for a specific function, rather than make it multipurpose. Figure 12-3 shows another gripper, designed for picking up Duplo sound modules.

Figure 12-3: A gripper for LEGO Duplo sound modules

the robotic arm

A *robotic arm* is a gripper mounted on a movable arm. When designing a robotic arm to lift a heavy load, we need to keep in mind that when you put a heavy load on a long arm, the motor powering that arm needs to be powerful enough to lift the load.

Since the NXT motors aren't terribly powerful, we'll use a counterweight to balance the arm and reduce the force needed to move the arm. When the counterweight is used, we only need to apply enough force to move the extra weight of the load, not to lift the arm itself. In the model of the crane robot shown in Figure 12-4, you see a Touch Sensor used as a counterweight.

A real tower crane uses several heavy metal or concrete plates to balance its load. In a large LEGO crane I once helped to build, two counterweights were used (as shown in Figure 12-5). One of the counterweights was fixed at the rear of the crane, and a second movable one was mounted along a gray track to increase or decrease the momentum on the arm. The two counterweights allowed us to balance the load on the crane.

Our CraneBot does not need a movable counterweight; the weight of the batteries and the NXT brick are sufficient to counter the weight of the ball.

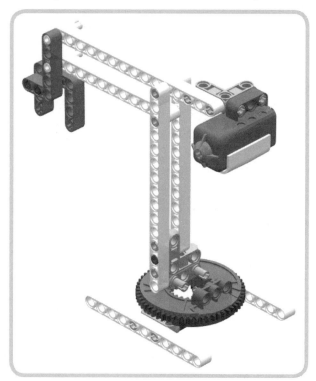

Figure 12-4: Model of a crane robot

Figure 12-5: A large LEGO tower crane

building CraneBot

Follow these instructions to build your own CraneBot! Start by building the gripper. At the end, you will use the longest wires in the NXT kit (50 cm/20 inch) to connect the Light Sensor to input port 4 and the motor to output port A. In order to guide the wires properly, we'll connect the cables later in step 37 of the body assembly.

building the gripper

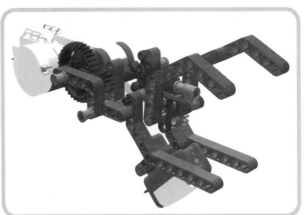

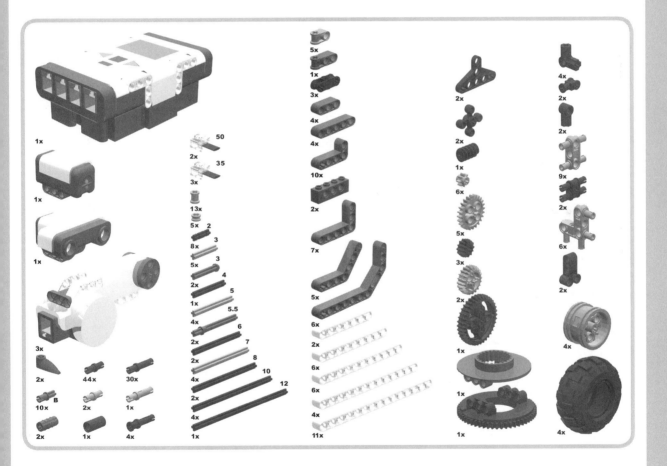

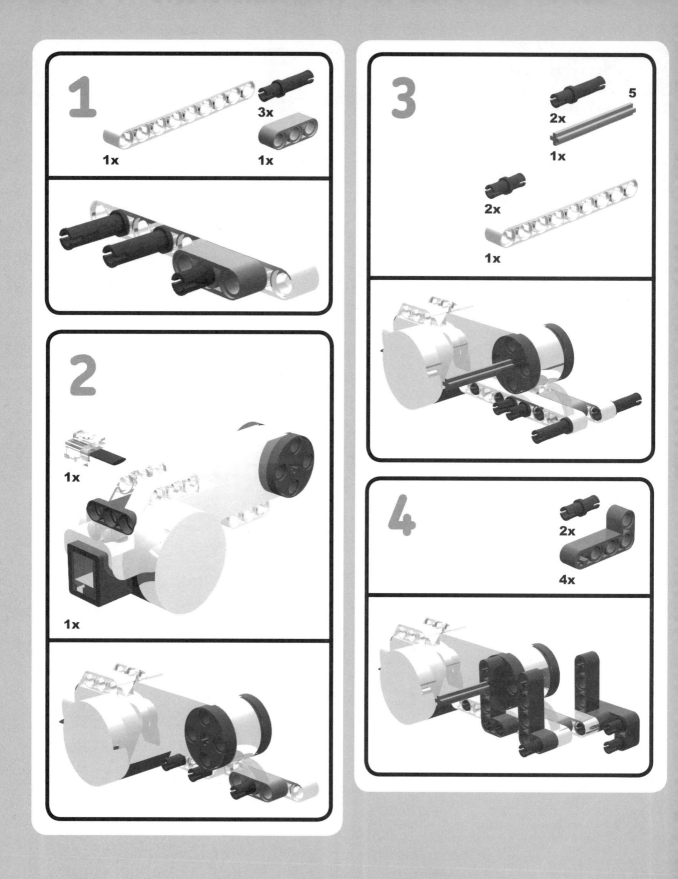

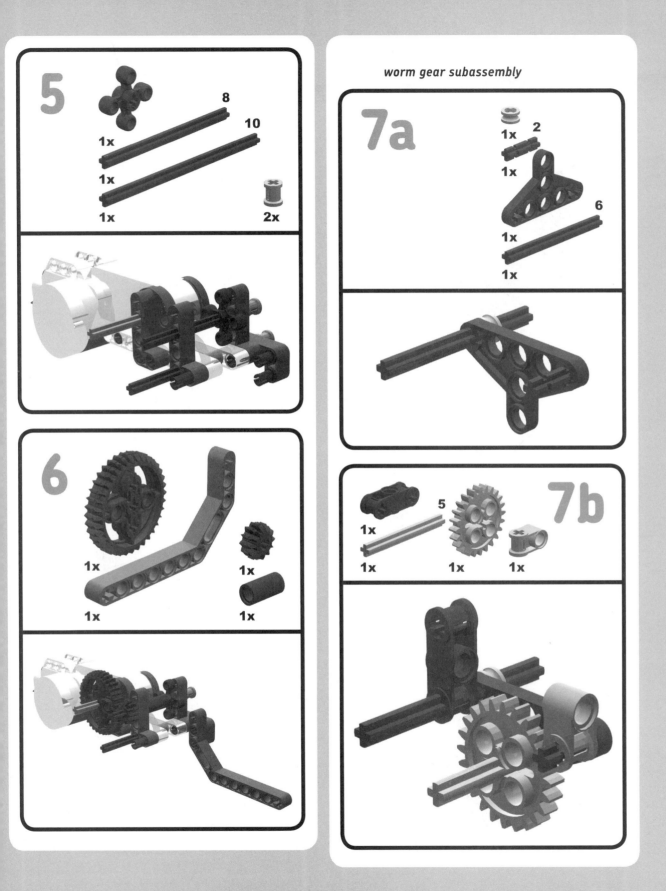

5

8
10

1x
1x
1x
2x

6

1x
1x
1x
1x

7a

1x
2
1x
6
1x
1x

7b

5
1x
1x
1x
1x

7c

1x   **6** **1x**   **1x**   **1x**

7d

1x
1x

Add the worm gear subassembly to the gripper. The length-6 axle goes into the axle hole of the gripper's double-bend beam.

8

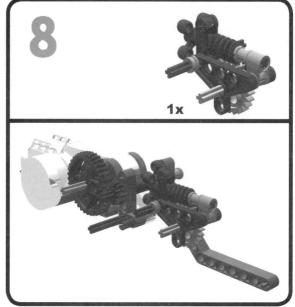

1x

light sensor subassembly using the 50 cm (20 inch) wire

9a

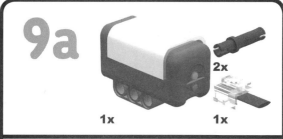

2x
1x   **1x**

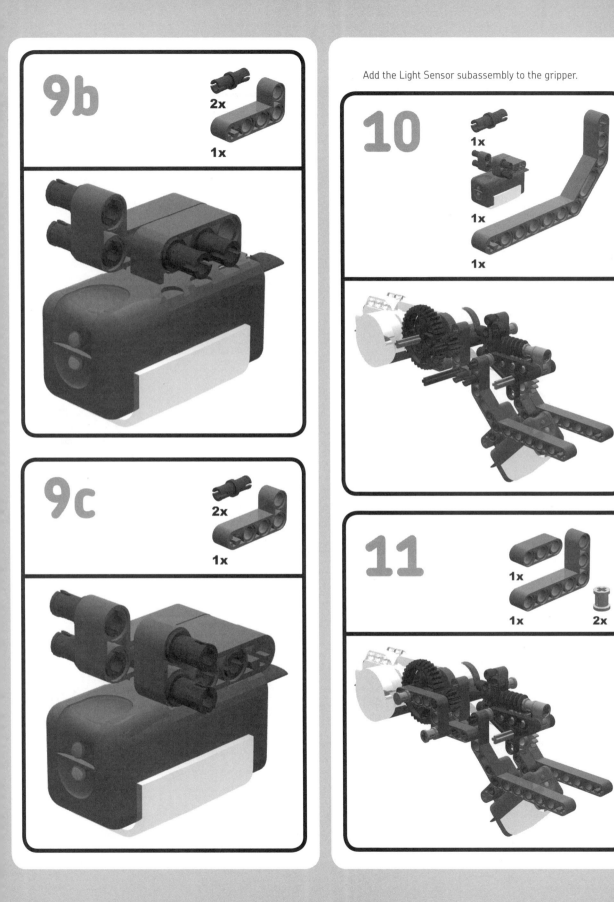

9b

2x
1x

9c

2x
1x

Add the Light Sensor subassembly to the gripper.

10

1x
1x
1x

11

1x
1x
2x

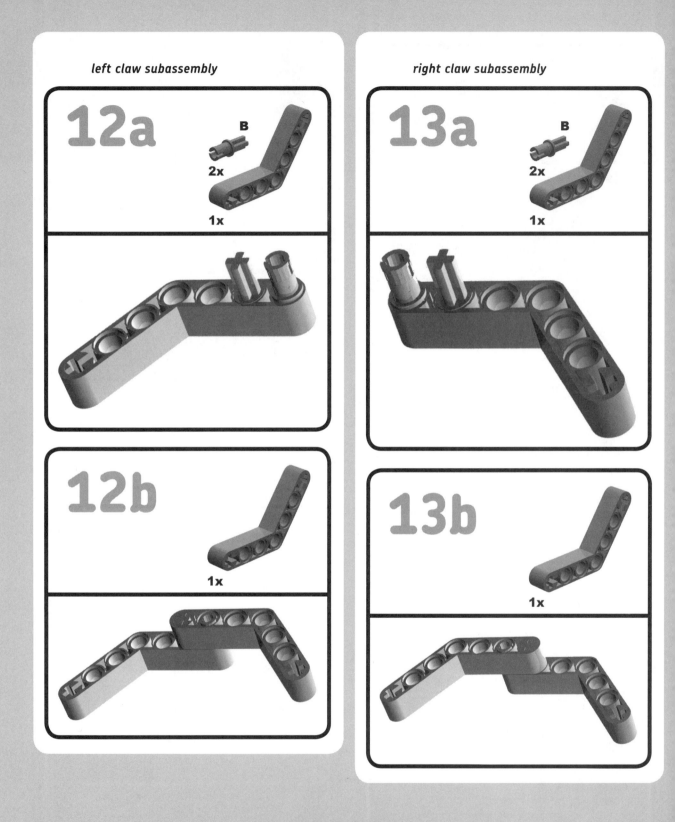

left claw subassembly

12a

B
2x
1x

12b

1x

right claw subassembly

13a

B
2x
1x

13b

1x

Add the claws to the length-5 gray axle on the gripper.

Test the gripper by turning the big 36-tooth gear wheel and grabbing a ball.

building the arm

The robot arm consists of two parallel beams that keep the gripper mounted at the front, horizontally, without the need for an extra motor to tilt it. (More motors would give the arm the ability to perform functions such as tilting and extending, thereby increasing the operating range of the robotic arm.) The CraneBot's arm can only be raised and lowered, using a lever at the back of the main body to move the arm up and down.

under arm subassembly

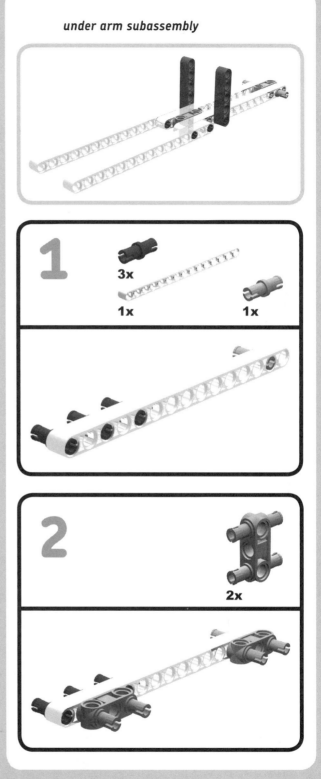

3

3x

1x   1x

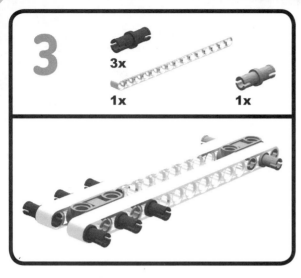

The dark gray, 5-hole beams between the upper and under arm are added to stabilize the arm, and they must be connected when adding the upper arm to the main body.

4

2x

2x

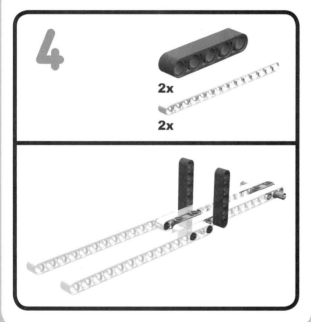

upper arm subassembly

Note the orientation of the NXT brick. This is important because of the limited cable length.

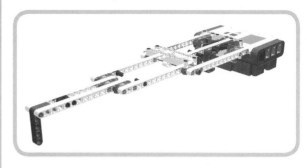

1

2x B

2x

2x

The counterweight (the NXT brick itself) is mounted only on the upper arm.

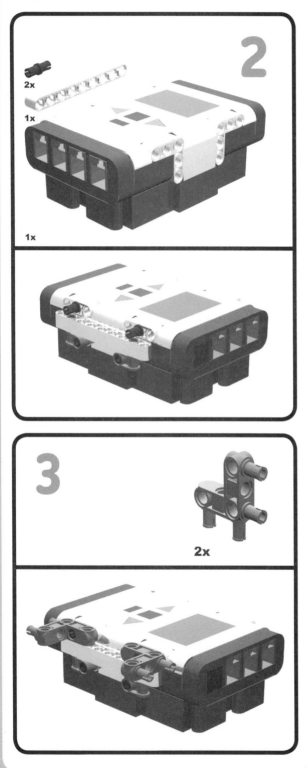

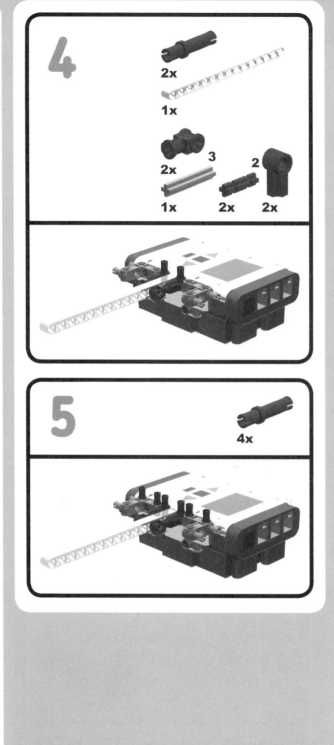

The black, length-3 friction pin is used as an end-stop. When connecting the upper arm to the body, we need to remove this axle and place it in the last cross-axle hole of the lever on the body.

The length-3 gray pin connects the gray, 3 × 5–hole, *L*-shaped beam to the 15-hole beams.

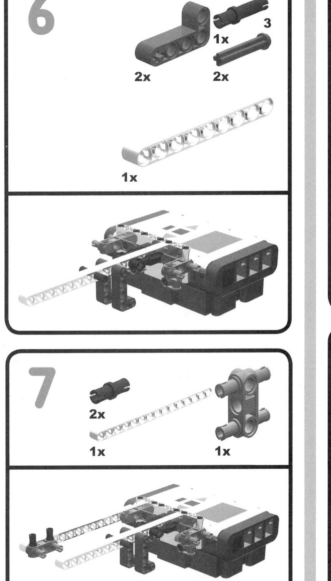

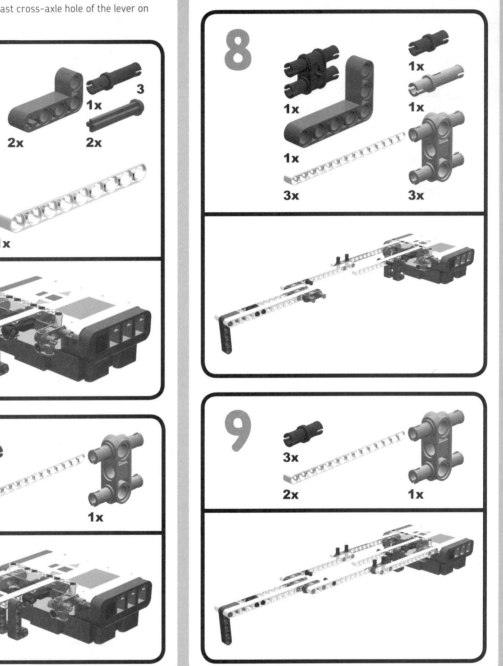

These two 9-hole beams make the arm sturdier, and the length-12 axle is used as the arms' main pivot point.

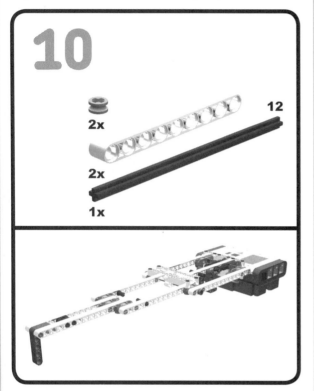

building the crane body

This robot uses wheels as a firm and stable foot. The wheels are placed on their sides for improved stability, but this makes the model unable to move around.

NOTE The foot subassembly is added as one of the last parts in order to keep the pictures clearer and easier to follow. You can build the foot subassembly first and add it on later, however, if you want to.

foot subassembly

Build four wheels to support the robot.

NOTE The side of the rim with the spokes is the underside; it faces the ground. The truck-like rim side is the inner side; it faces up.

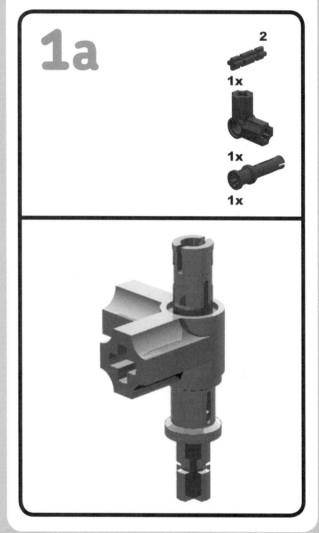

Build two of these side frames.

Add the previously built frames to the turntable with the black side facing down.

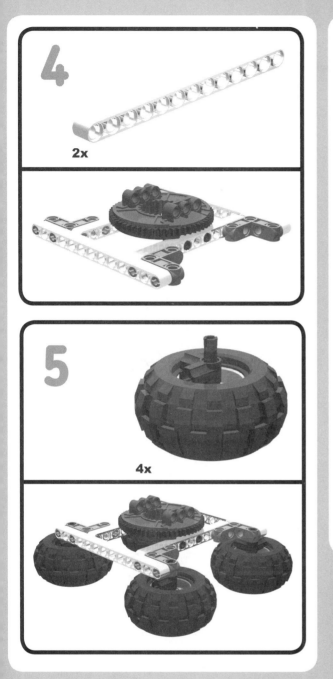

main body subassembly

With a stable foot in place, it's time to build the main body.

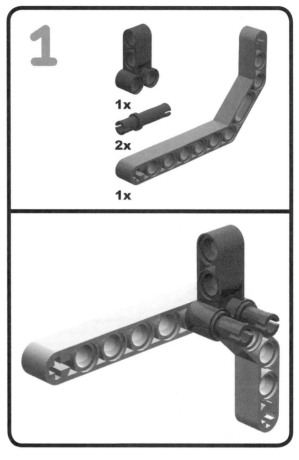

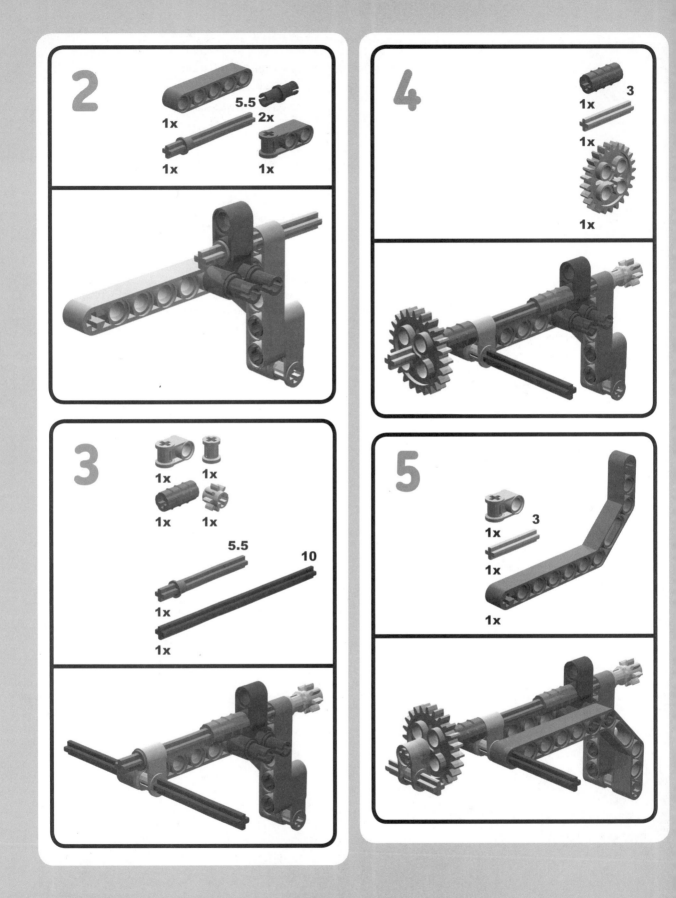

Add the foot (only the turntable is shown here) by placing the length-8 axle and pushing in the length-3 friction pins.

Start with the right side of the body.

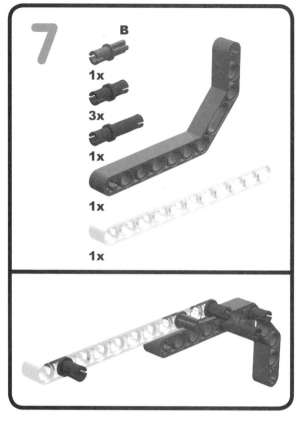

8

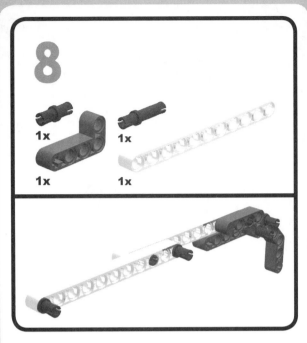

1x   1x

1x   1x

Put a length-2 axle in a 20-tooth double-bevel gear, and place it in the motor.

9

2x   1x   1x

2

1x

1x   1x

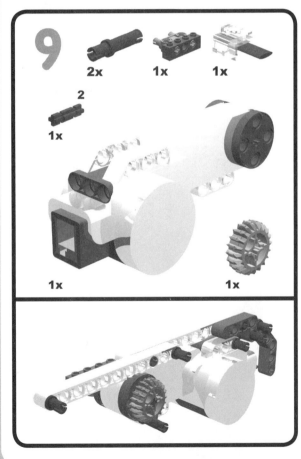

Put the bushings on the axles. (They are not shown in the picture since they are on the back side.)

10

1x

10

1x

2x

1x   2x

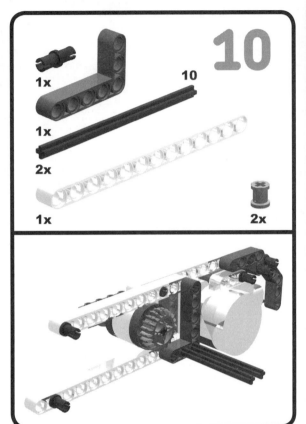

Build the turning subassembly.

11a

4

1x

1x   1x

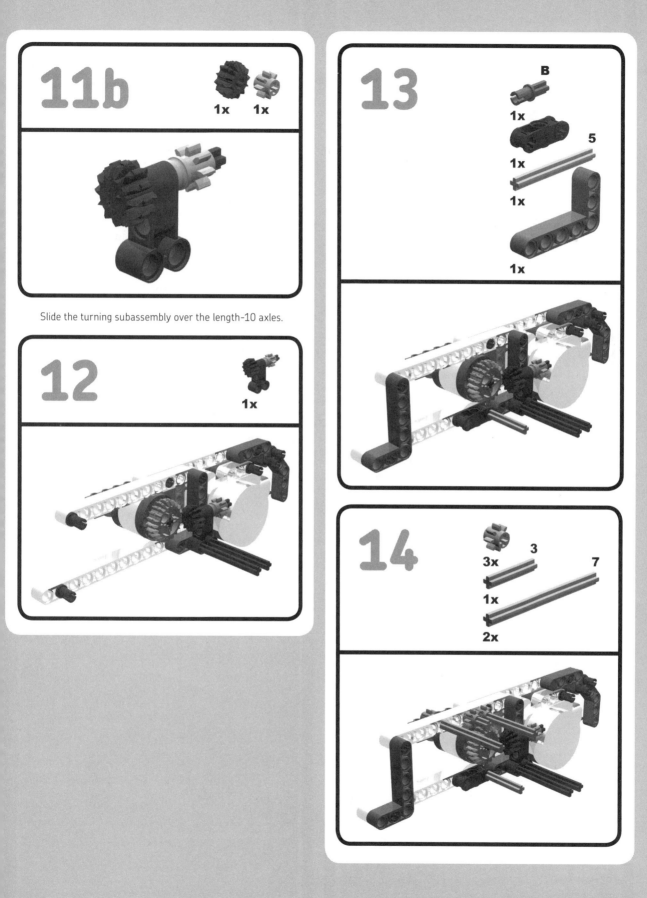

11b

1x 1x

Slide the turning subassembly over the length-10 axles.

12

1x

13

B

1x

1x   5

1x

1x

14

3x   3

1x   7

2x

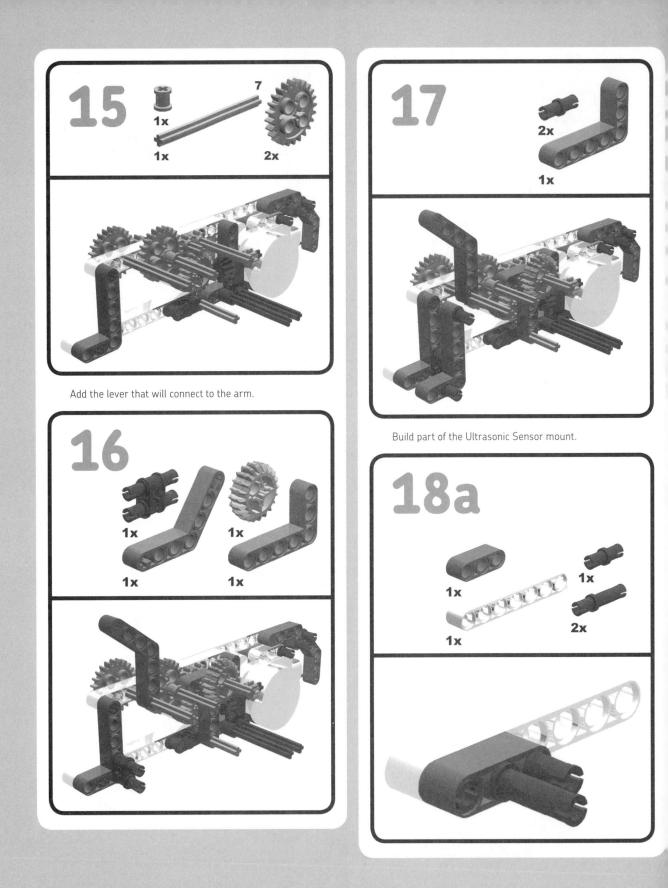

15

1x
1x
7
2x

Add the lever that will connect to the arm.

16

1x
1x
1x
1x

17

2x
1x

Build part of the Ultrasonic Sensor mount.

18a

1x
1x
1x
2x

Use a 35 cm (15 inch) cable here.

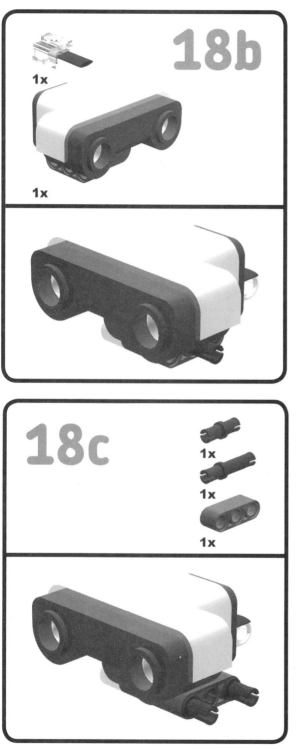

Mount the Ultrasonic Sensor on the last friction pin. As shown in these photos, the sensor should face slightly downward with the motor's cable now guided upward and placed between the pins.

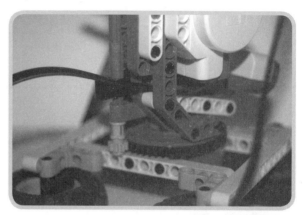

Carefully combine the middle and right section of the main body. Be careful when handling the length-3 axle with the perpendicular gray axle connector, because it might fall off while mounting the body parts. (Note that the Ultrasonic Sensor is not shown in the drawings.)

20

1x

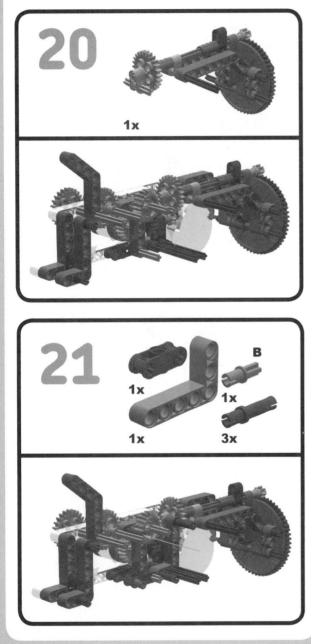

21

1x

1x

B

1x

3x

22

1x

1x

Attach the under arm with the gray pins to the body in the fifth hole from the top of the white beam.

23

1x

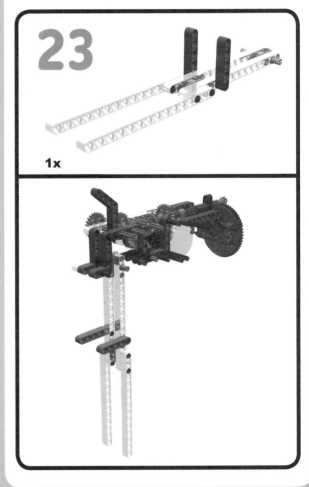

24

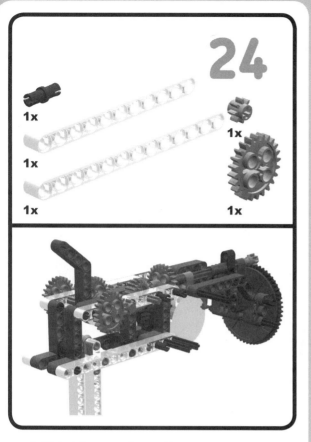

1x

1x

1x

1x

1x

Build the left motor subassembly.

25

1x

B

1x

1x

1x

1x

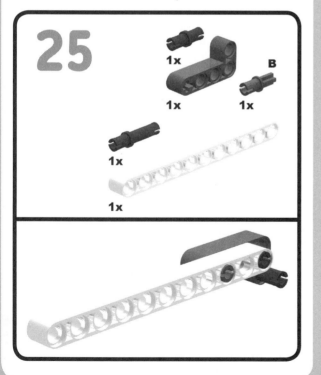

26

1x

1x

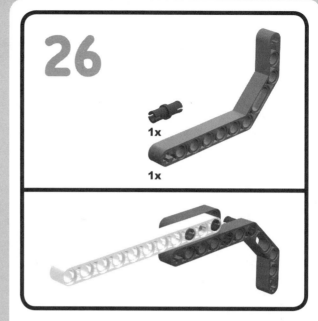

Put the cable (35 cm/15 inch) in the motor and attach the motor to the assembly, then place a length-3 friction pin (see arrow). Although the arrow goes all the way through, the pin does not. (The arrow is pointing to the hole that the pin should slide in from the back side of that beam.)

NOTE In this drawing the Ultrasonic Sensor is included. Make sure that its cable is placed between the *L*-shaped beam and the 7-hole white beam of the Ultrasonic Sensor.

27

1x

1x

1x

3

1x

1x

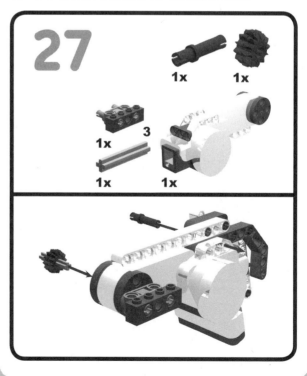

Combine the left motor subassembly with the main body. It might take a little force to push all the axles and pins into the correct holes. Be sure that the motor cables are facing up.

28

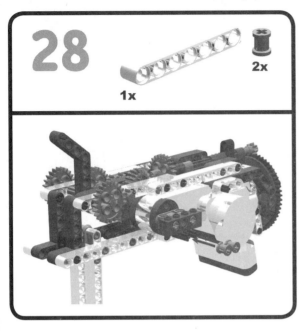

1x   2x

29

In order to add the upper arm to the body, we need to remove the length 3 axle with stud as well as the blue axle pin.

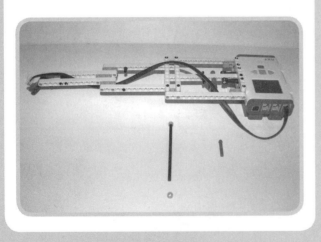

30

Turn the lever used for lowering the arm level just a little bit.

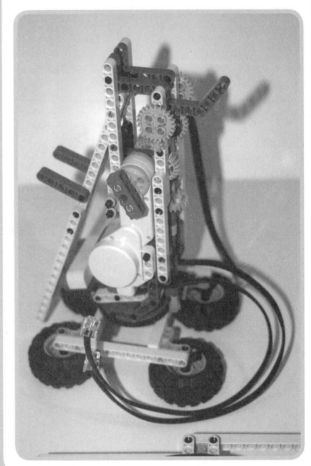

31

Connect the upper arm using the length-3 studded axle.

32

Place the upper arm on top of the body and slide in the length-12 axle and half-bushing to prevent it from becoming loose.

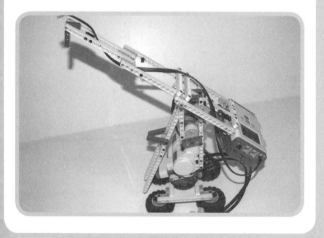

33

Connect the gray, 5-hole beams to reduce the flex in the arm. This will connect the under arm with the upper arm.

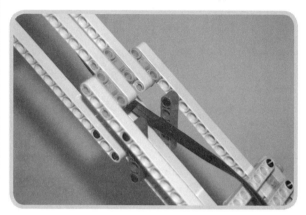

34

Use length-5 and length-7 axles to mount the gripper.

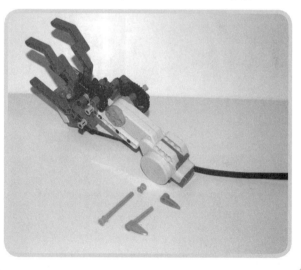

35

Place the length-7 axle through the motor and the dark gray, *L*-shaped beam.

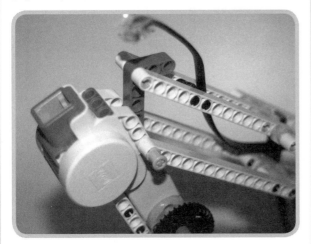

36

Place the length-5 axle.

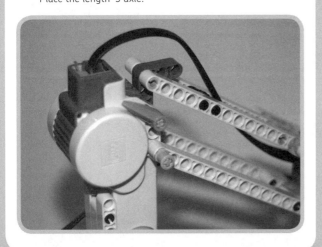

37

Connect all the wires to the NXT according to Table 12-1. For additional guidance, see Figure 12-1 on page 143.

table 12-1: attaching the motors and sensors

motor/sensor	wire length (cm)	wire length (in)	NXT port
Gripper motor	50	20	Output A
Right turning motor	35	15	Output B
Left lifting motor	35	15	Output C
Ultrasonic Sensor	35	15	Input 3
Light Sensor	50	20	Input 4

programming CraneBot

I'll provide a simple CraneBot program in this book. You'll find other, more sophisticated programs on this book's companion website.

The following program rotates CraneBot and scans with the Ultrasonic Sensor for a ball. Once CraneBot detects the ball, the arm will move down and grab it. The Light Sensor will be used to determine the color of the ball and will display the color on the NXT's LCD.

1. Build the ball holder used on the TriBot, as outlined in the manual or NXT-G software for the NXT Retail version (beginning on page 60).

2. Make sure that the movable fingers on the gripper are open wide enough to go over the ball, then lower the arm by turning on the left motor until the gripper is just above the ground.

3. Place the ball in the ball holder, then place this assembly underneath the gripper to be sure that CraneBot can see and detect the ball.

4. Move the arm up to its highest position and rotate it about one quarter turn clockwise, to the right (as seen from above). This will be the starting position for the program that follows.

The program is divided into four distinct parts. The flow of the program goes like this:

1. Find the ball.

2. Grab the ball.

3. Check and display the color of the ball.

4. Wait for the user to read the display, then end.

part 1

Part 1 of the program, shown in Figures 12-6 and 12-7, finds the ball. In this part, CraneBot loops forever and turns clockwise (motor B) while it scans with the Ultrasonic Sensor for objects closer than 8 inches (20 cm).

Figure 12-6: Start the loop.

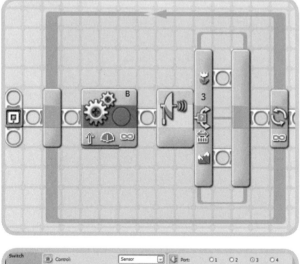

Figure 12-7: Add Move and Switch blocks for the Ultrasonic Sensor.

The program steps in Figure 12-8 will be added to the upper (near) branch. (Note that the Sound block does not wait for completion, and you can play any tone you like.)

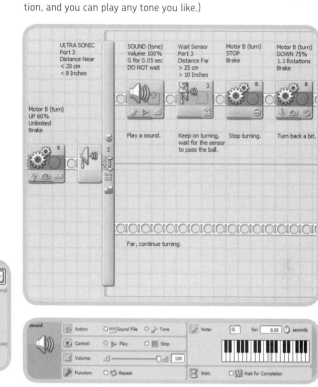

Figure 12-8: Add the Sound, Wait for Ultrasonic Sensor, Brake motor B, and Turn Back motor B blocks.

Because the Ultrasonic Sensor is not well suited for detecting round objects, CraneBot will continue to turn until the Ultrasonic Sensor no longer detects the ball. Once the sensor detects the ball, CraneBot stops turning then turns backward for 1.1 rotations of the motor.

part 2

Here we grab the ball. As shown in Figure 12-9, CraneBot lowers the arm (motor C) for 3.5 rotations and then closes the gripper. The gripper (motor A) will run at full speed (100 percent) for one second.

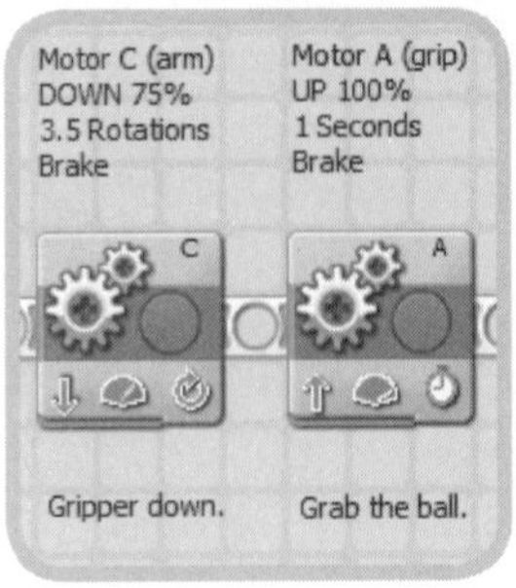

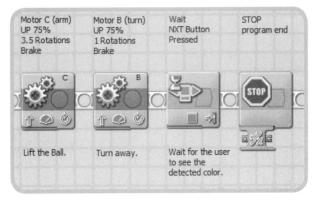

Figure 12-9: Lower the arm and close the gripper.

part 3

In this part we check and display the color of the ball (Figure 12-10). Use the Light Sensor on port 4 to look at the color of the ball. (Remember to calibrate the sensor first.) When the color of the ball is less than 40, the ball is blue.

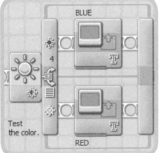

part 4

As shown in Figure 12-11, in this part we move away from the ball holder. We first move the arm up (3.5 rotations) then rotate it clockwise one time. Next, the program waits for the user to look at the display for the text *Blue* or *Red*. Pressing the orange Enter button on the NXT stops the program.

Use your imagination to extend this program by adding a second ball holder that would allow the gripper to drop the ball and continue looking for a new one.

Figure 12-11: Move the arm up and turn clockwise, then wait for the NXT button to be pressed and stop.

Figure 12-10: Check and display the color of the ball.

slot machine:
a one-armed robot

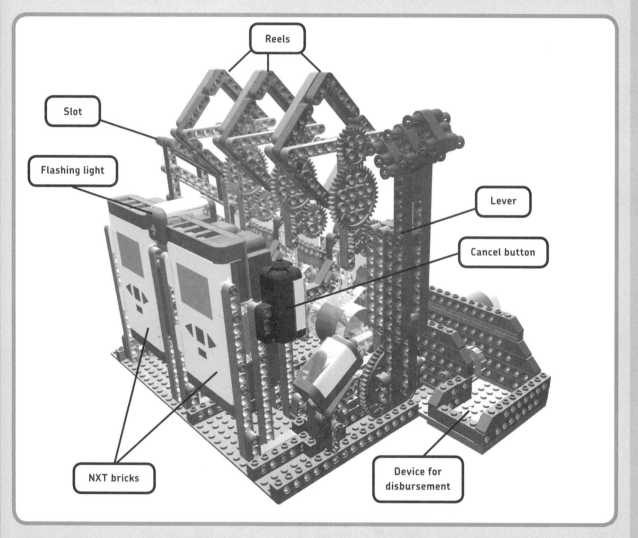

Figure 13-1: The Slot Machine

what is a slot machine?

Today, *slot machines* (also called *poker machines* in Australia or *fruit machines* in the United Kingdom) are the most popular type of gambling machines in casinos. In their traditional form, they principally consist of a set of reels (usually three) that spin once a lever on the side of the machine is pulled. To activate the game, a single player inserts coins into a slot somewhere in the upper front part of the machine—hence the name *slot machine*.

If, when the reels stop spinning, the symbols on the reels form a particular pattern in the viewing window (called the *payout line*), the player wins.

NOTE Because the lever on a slot machine makes it look a bit like a one-armed man (you may need to stretch your imagination), they are also known as *one-armed bandits*. Some people say that this name refers to the slot machine's ability to completely rob the gamer of her money—something they seem to do quite well, as slot machines today contribute 60 to 70 percent of revenue to United States casinos.

While classic slot machines were purely mechanical, modern electronics and computers have given us new forms. Some types of modern machines, for instance, referred to as *ticket in, ticket out machines*, accept only a bar-coded stripe of paper (a "ticket") instead of coins, and they pay winnings with a similar stripe that the gambler may cash in later. Other changes include completely computer-based controls (in particular, wins based on a random number generator), screen-based display and input, remote connection to user credit accounts, and the replacement of the lever by a button (because the kinetic energy formerly provided by the pull of the machine's arm is no longer required to work the mechanical components).

hardware challenges

To create a slot machine with LEGO MINDSTORMS NXT that still resembles a classic model (see Figure 13-1), our major hardware challenges boil down to implementing the following components:

* Reels
* Slot
* Pay-off mechanism
* Lever
* Store

the reels

I chose to implement the reels as square components rather than round ones, because square components are easier to build with LEGO and simpler to control. You can see an example of a reel in Figure 13-2.

the lever

The lever has to be reliable as well as robust and stable, so we use an assembly of beams attached to a vertical turntable, as you can see in Figure 13-3. The turntable prevents the lever from swinging left or right while also allowing for back-and-forth mobility. A Touch Sensor manages the detection of pulls.

the slot

The slot must reliably detect when coins are inserted and make sure that they are transported to the internal coin store (which is implemented as a chain of treads) without losing any coins.

For detection, we use a Light Sensor, which detects the coins as they run from the slot to the store (i.e., the warehouse for the coins that the player has inserted). The Light Sensor detects the coin as it passes through the slot and reflects the beam of light sent out by the sensor. Figure 13-4 shows the slot and the Light Sensor.

WORTH KNOWING

The precursors of the modern slot machine were tightly connected to classic multi-player casino games, such as poker. In 1891 two engineers in Brooklyn, New York, developed a machine that mimicked a poker player by providing 50 card faces on five drums. This machine proved extremely popular in bars all over New York, but, due to the rather complicated poker rules, an automatic pay-off from the machine was virtually impossible, at least by pure mechanical means. Before long, the cards were replaced by symbols on spinning reels; the first machine of this type was invented by the German immigrant Charles Fey in San Francisco. Fey's machine used the Liberty Bell as one of its symbols—hence the origin of the bell symbol on slot machines. Another popular early machine paid off in flavored chewing gum while displaying images of the flavors, like cherries or melons, as symbols. These, as well as BAR (the logo of the company producing the chewing gum, Bell-Fruit Gum Company), and the Liberty Bell have survived on slot machines until today.

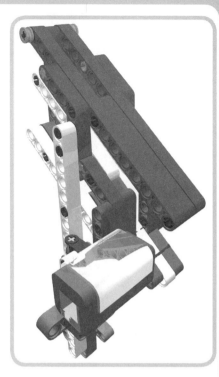

Figure 13-2: The reel

Figure 13-3: The lever

Figure 13-4: The slot with the associated Light Sensor

the pay-off mechanism

The task of the pay-off mechanism is to transport a certain number of coins specified by the controlling program from the store to the device that disburses the coins the player has won. We use a chain of treads both for the store and the pay-off, as you can see in Figure 13-5.

Coins are stored on the treads; when a coin needs to be paid off, the treads are moved forward until the leading coin falls into the output device.

the complete robot

Since this robot uses five motors, we'll need two NXT bricks. By placing them on the front of the Slot Machine, we can use their LCDs as screens for the Slot Machine to display the results of each spin.

Figure 13-5: The combined store/pay-off mechanism

Figures 13-6 and 13-7 show the basic setup with the two NXT bricks as well as the Slot Machine from the back.

I've added a Light Sensor to provide some flashy light effects to the machine (see Figure 13-8). To do this, we do not read the sensor's values but use its ability to send out light with its built-in beam, making it flash when the reels are spinning or the player has won a game.

Additionally, a Touch Sensor acts as a Cancel button (Figure 13-9), giving the gambler the ability to cancel the whole game and get back any money that has yet to be used.

Figure 13-6: The Slot Machine, built with LEGO

Figure 13-7: The Slot Machine (viewed from behind)

Figure 13-8: The flashing light

Figure 13-9: The Cancel button

building
the slot machine

Unfortunately, you won't be able to build this model with only one
NXT kit. But it's so much fun, it's worth getting two!

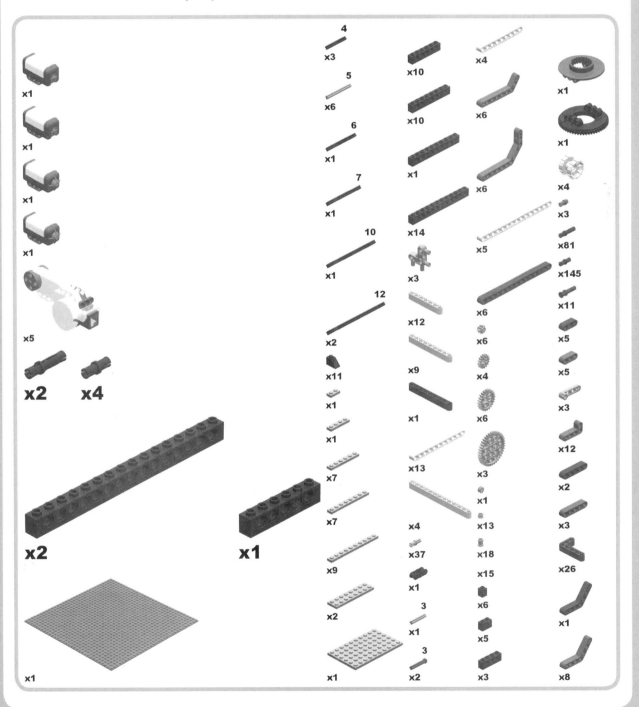

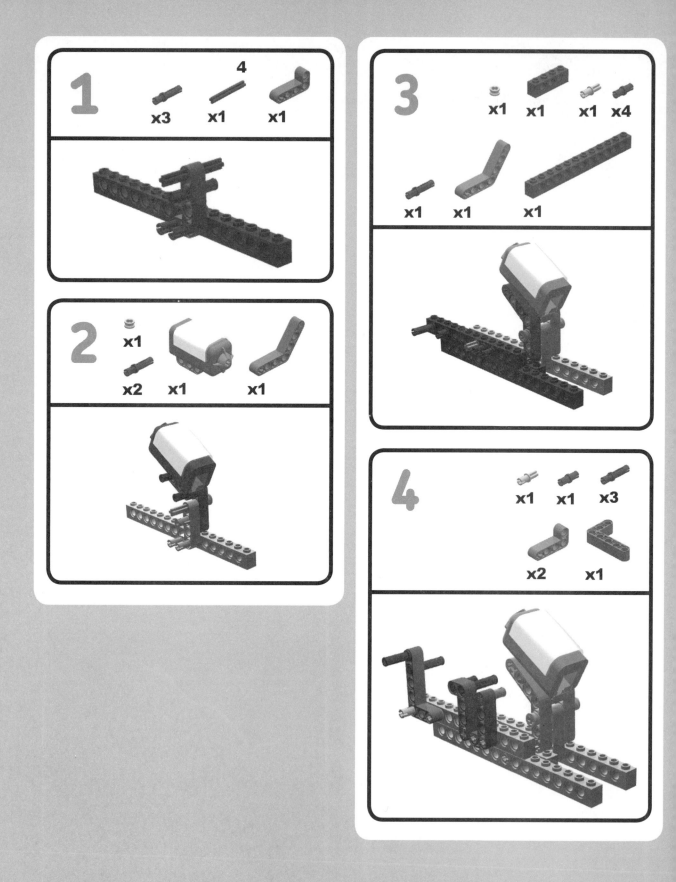

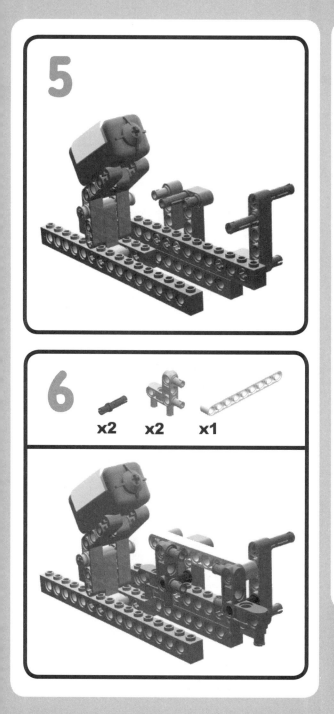

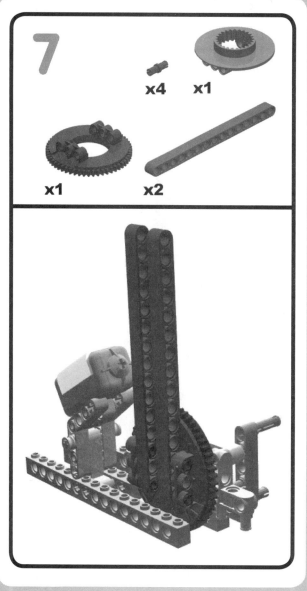

8

x2 x6 x1 x2 **3**

9

x1 **3** x3 x2

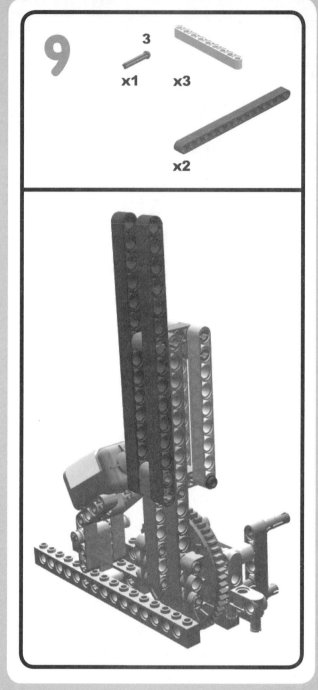

10 x2 x1

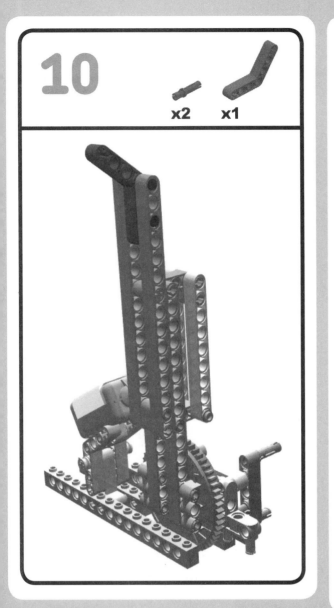

11 x11 x5 x2

12

13

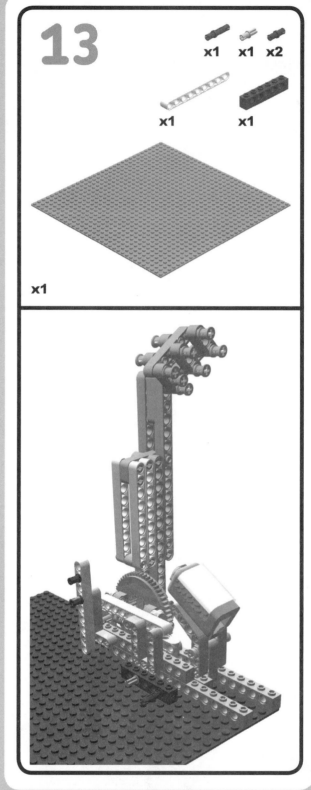

14

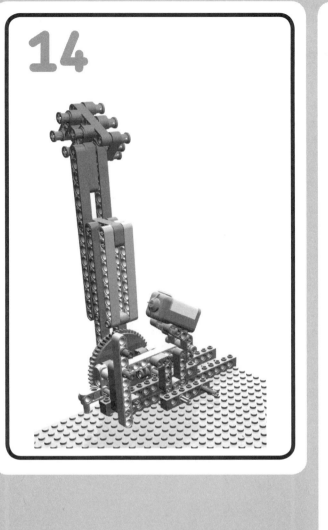

15

x3 x1

x1

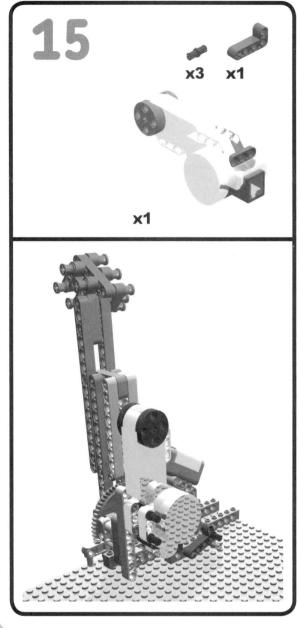

16

x3 x2

x1 x1

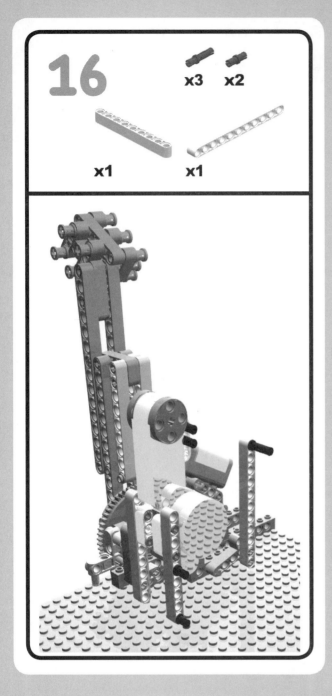

17

18

x1   x3

x1   x2   x1

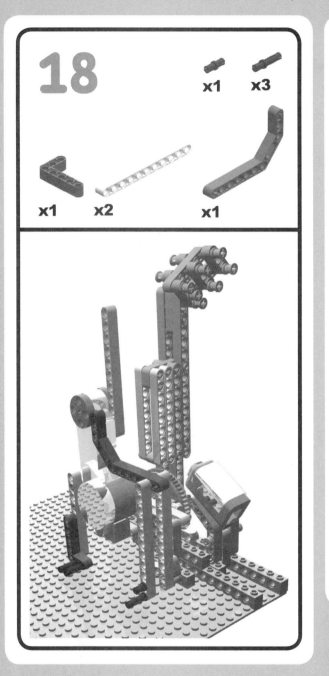

19

x1   x1   x1   x1

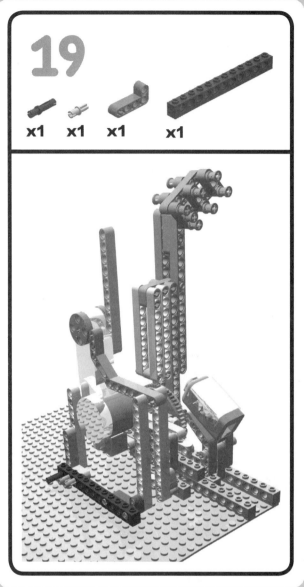

20

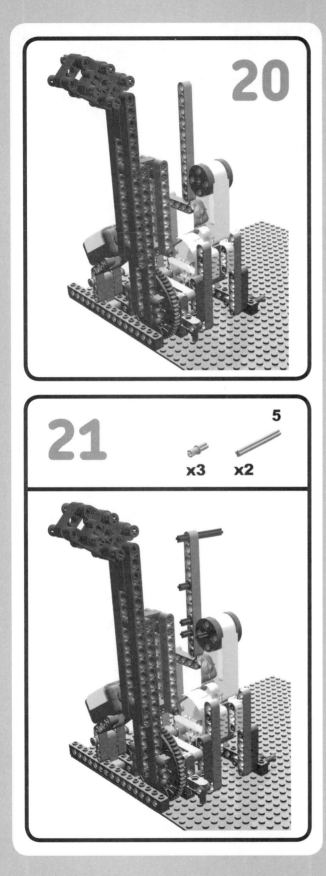

21 5
x3 x2

22
x2 x2 x1

23

24 x1 x1 x1 x1

x2 x1 x1

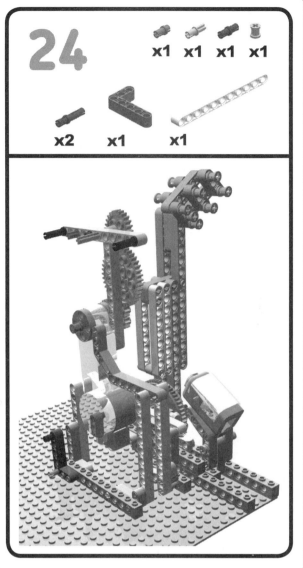

25

x1 x6 x1 x2

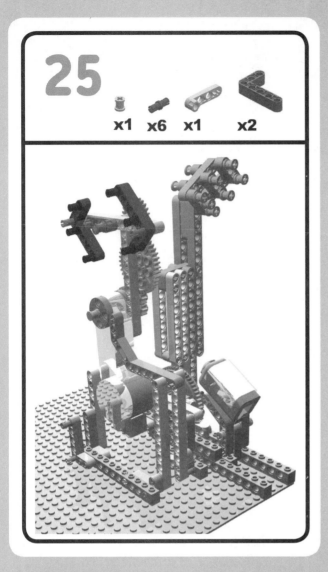

26

x8 x4

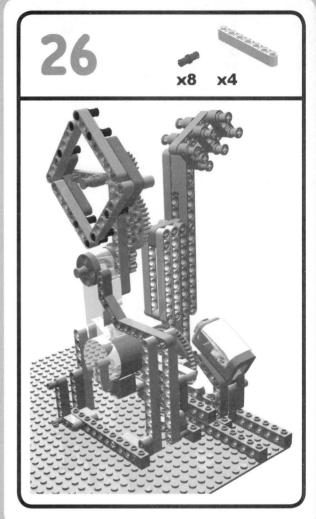

27

x2

28

x2

x1 x1 x1

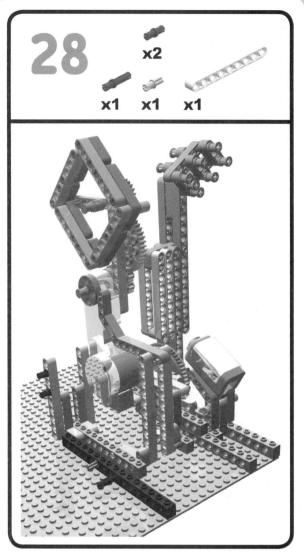

29

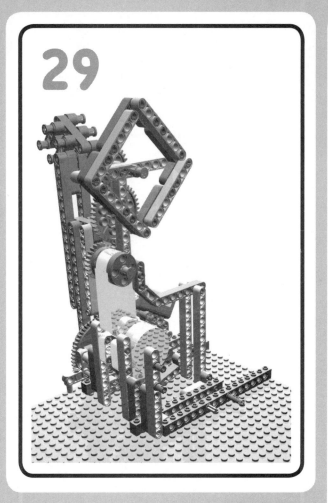

30

x3   x1   x1

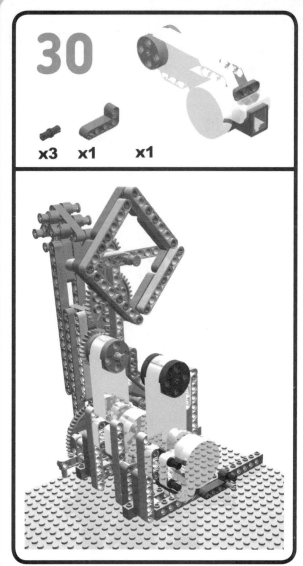

31

x3 x2

x1   x1

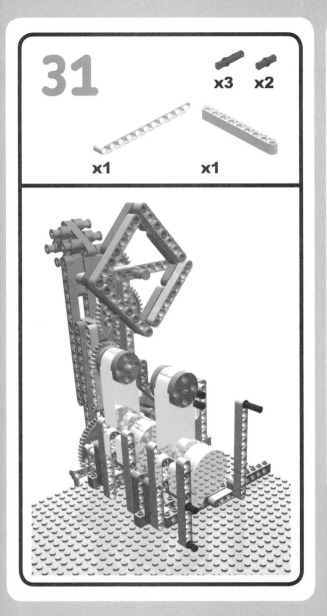

32

33

x1 x3

x1 x2 x1

34

x1 x1 x1

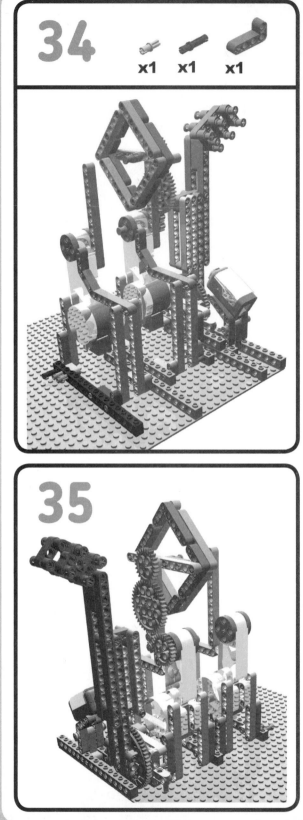

35

36

x3 x2 5

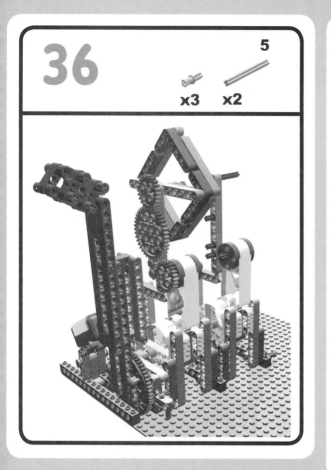

37

x2 x2 x1

38

39

x1 x1 x1

x1 x2 x1 x1

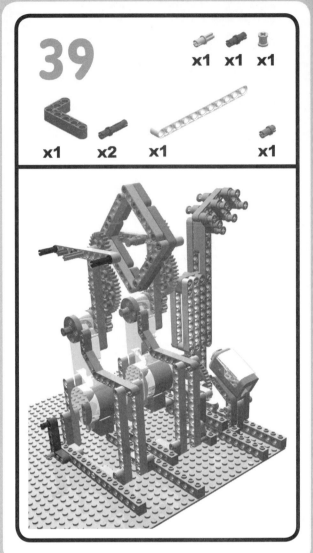

40

x1 x6 x1 x2

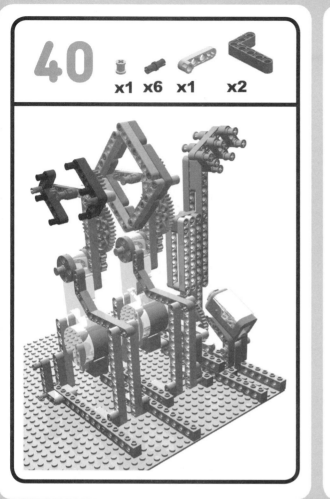

41

x8 x4

42

x2

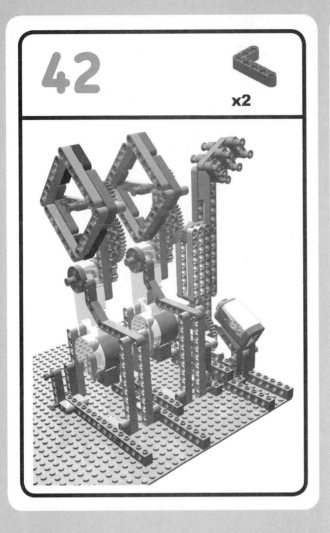

43

x1 x2 x1

x1 x1

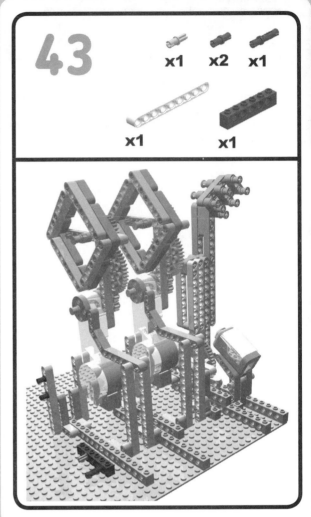

44

45

x3 x1 x1

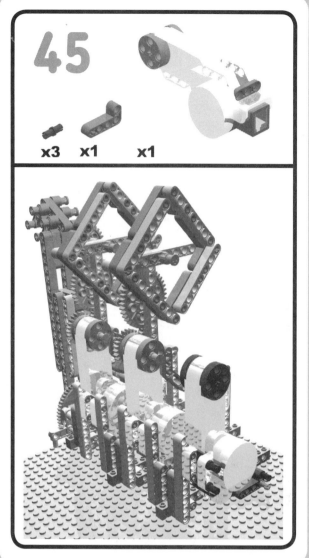

46

x3 x2

x1 x1

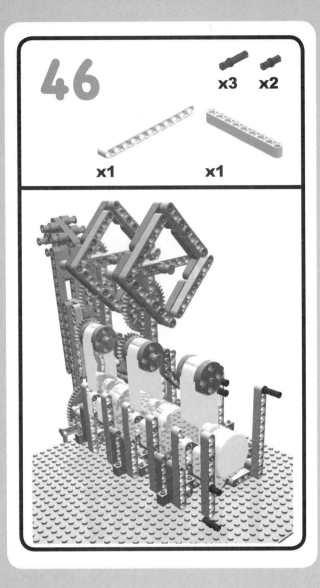

47

48

x1 x3

x1 x2 x1

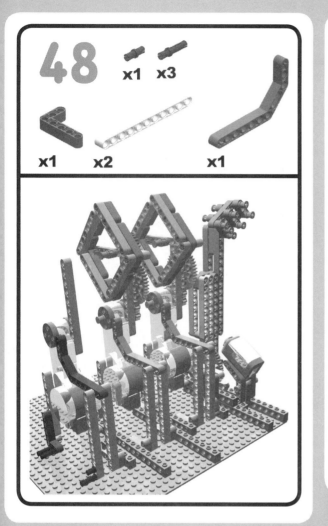

49

x1 x1 x1 x1

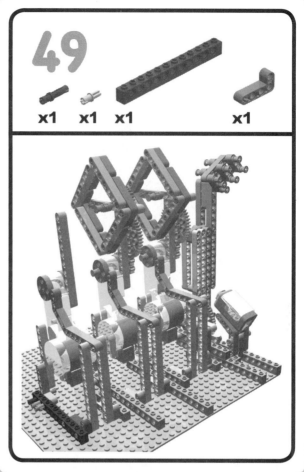

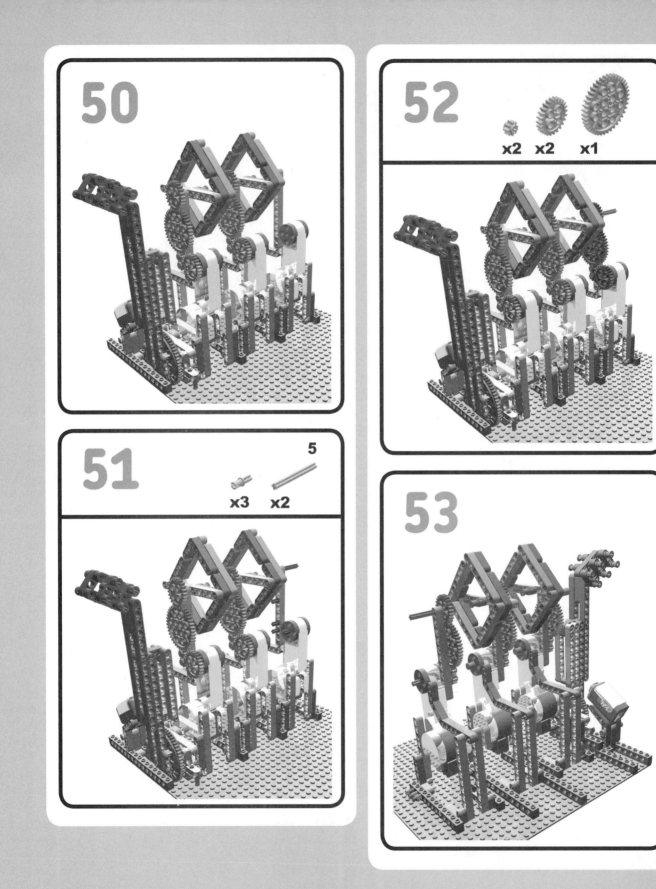

50

51

5

x3   x2

52

x2   x2   x1

53

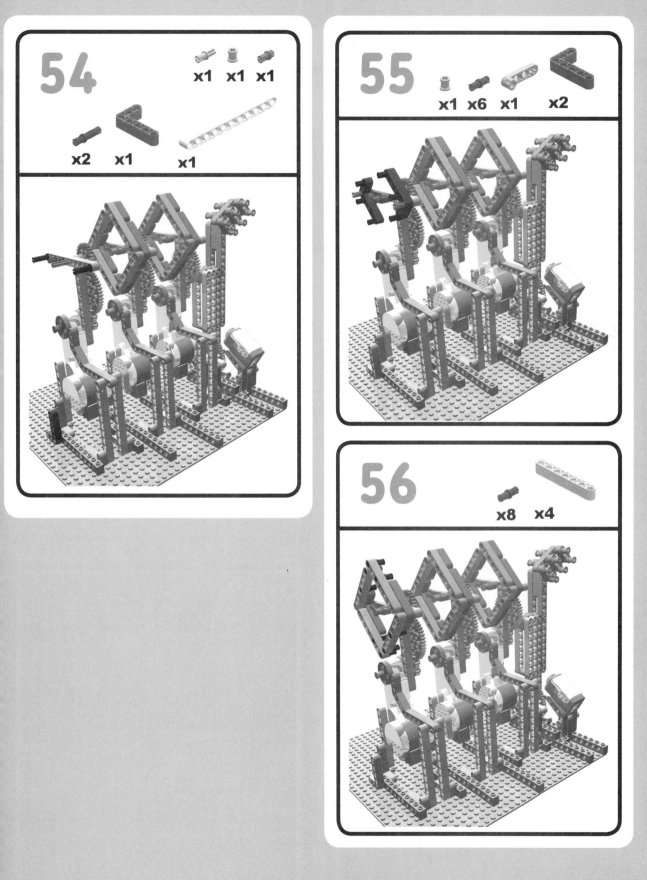

57 x2

58 x1 x3

59 x1 x1 x2 x2 x1

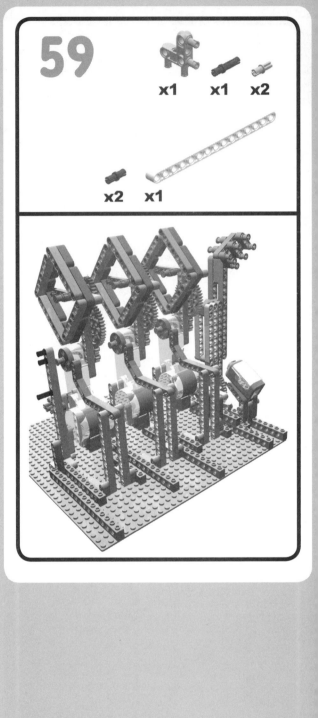

60

x1 x1 x1 x1

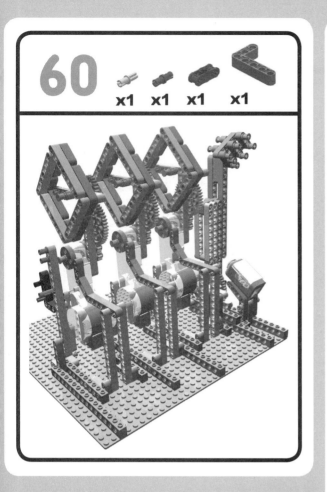

61

x3 x3

x1 x2

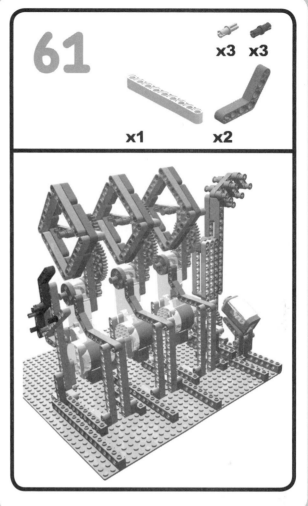

62

x2   x1   x1

x1   x1   x1

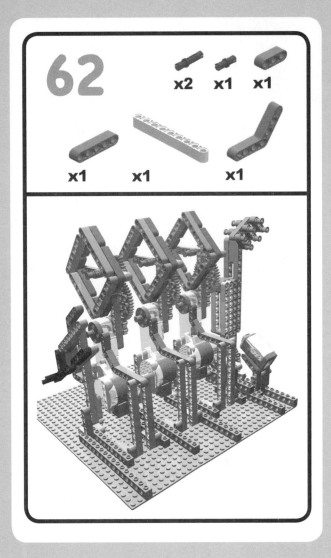

63

3

x1   x1   x6   x1

64

x1   x2   x1   **4**

x1   x1

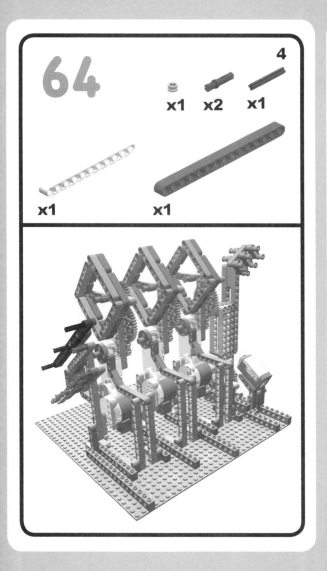

65

x1

x1   x1

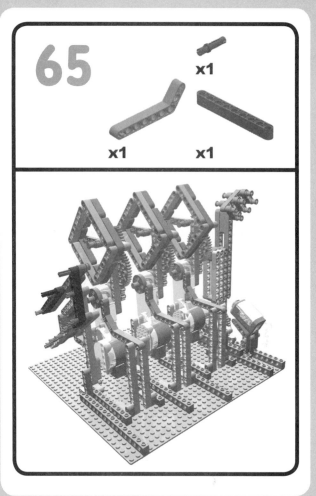

66

x1
x2
x1
x5
x1
x1
x1

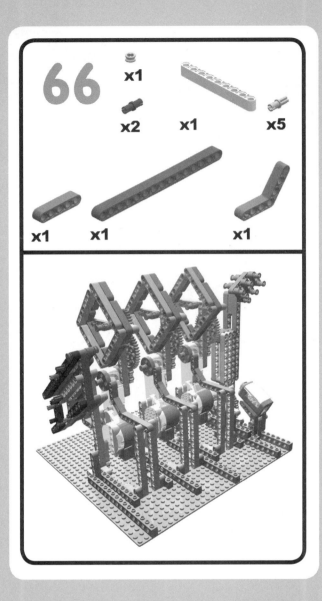

67

x1 x1 x1 x2

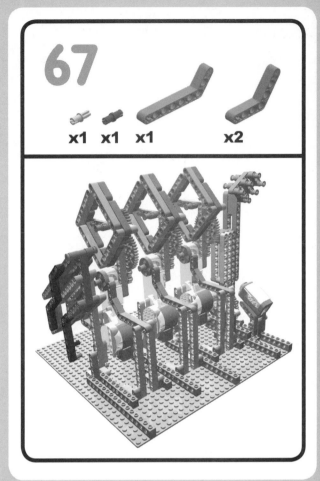

68

x3   x2

x1

69

70

x1   x1   x1

71

x1

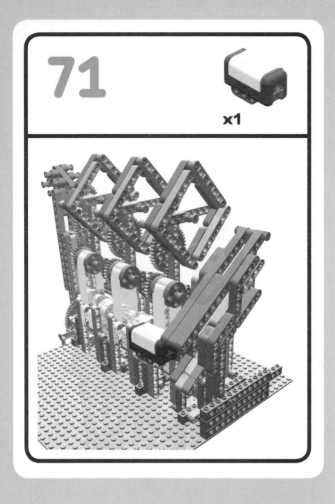

72

x1 x4 x1

x2 x3

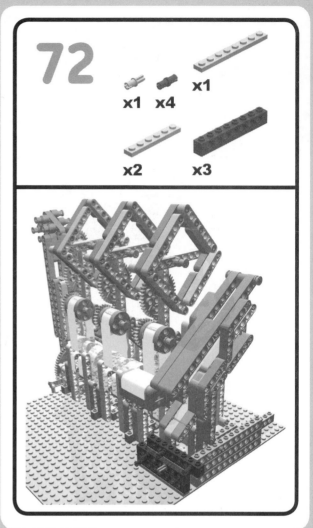

73

x2

x1

6 x1

x1

x1 x2

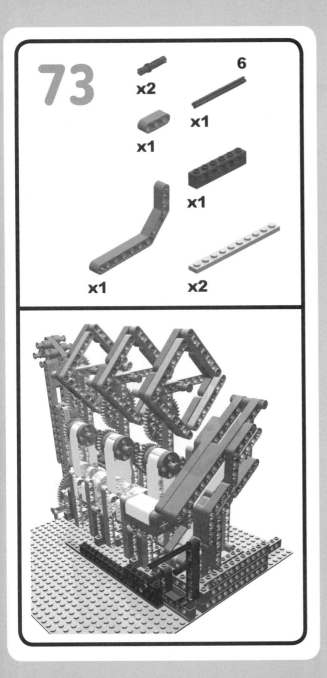

74

x2

4

x4 x1

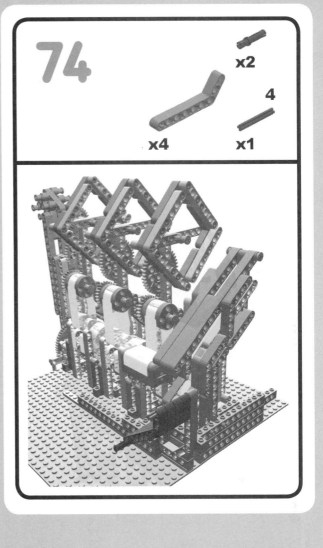

75

x2 x1

x1

10

x1

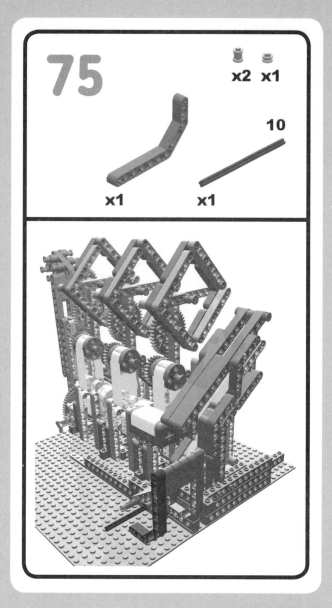

76

x1

x1

x1

x1

x2

x1

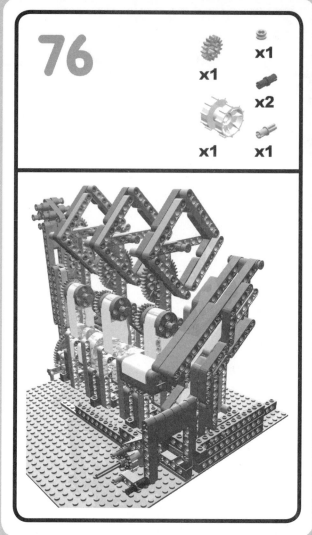

77

x2

x2   x1

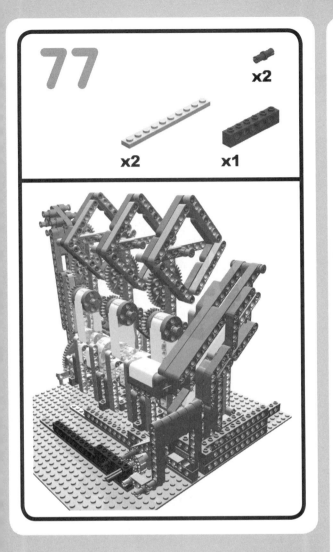

78

x2

x3   x1

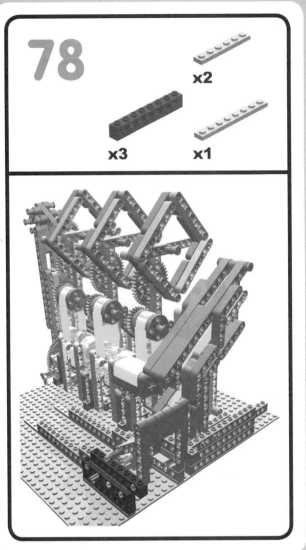

79

x1
x1 x2

80

x1
x3
x2 x1

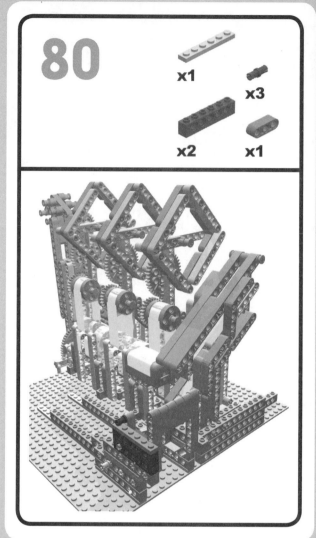

81

x2

82

83

x1   x1   x3

x2   x1

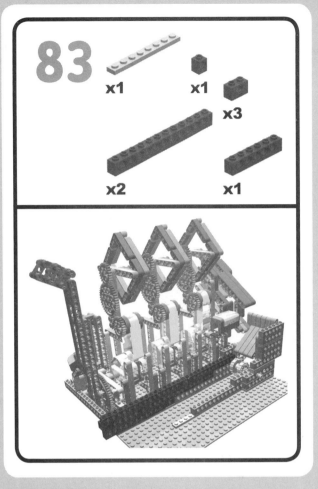

84

x1

x1

x1

x1

x2

85

12

x1

x1

x1

x2

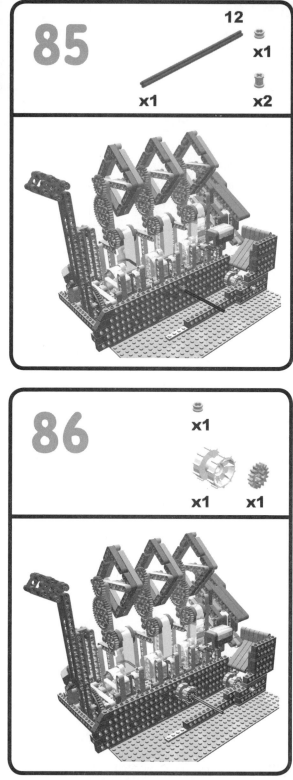

86

x1

x1

x1

87

12
x1
x1
x2

88

x1
x1
x1

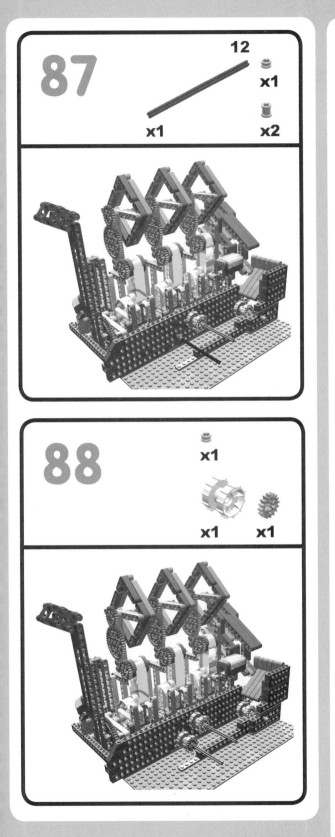

89

7
x1 x1 x1 x1

90

x1 x1 x1

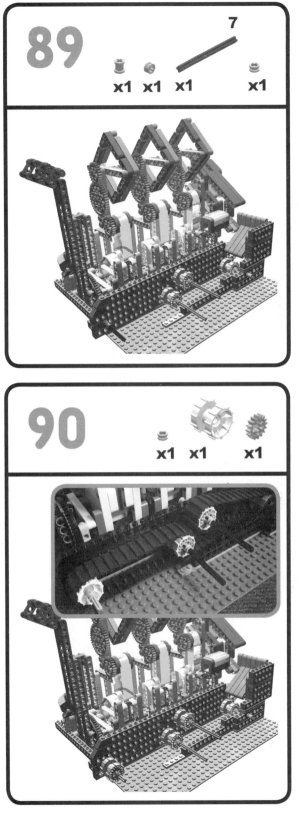

91

x1 x2 x1 x1

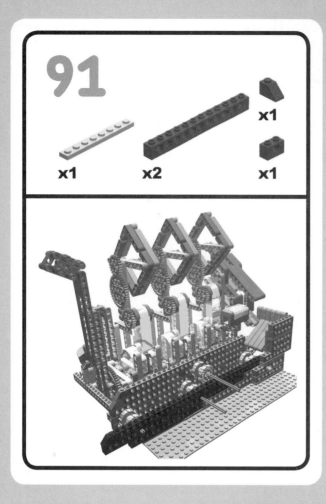

92

x1 x1 x5 x1 x1 x1 x1 x2

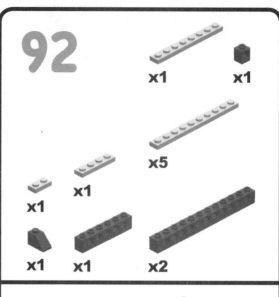

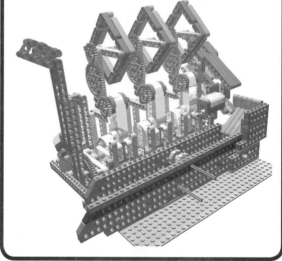

93

x1 x1 x1

x2 x1 x1

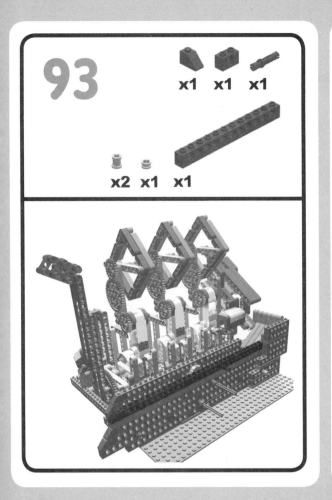

94

x2

x2

x1 x1

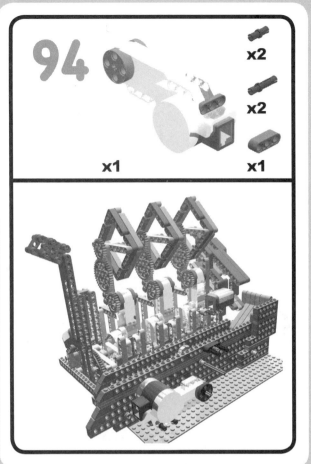

95

x1 x1

x2 x1

x1 x1

96

x1 x1

x1

x1 x1

97

98

x1

x1

x2

x2

x2

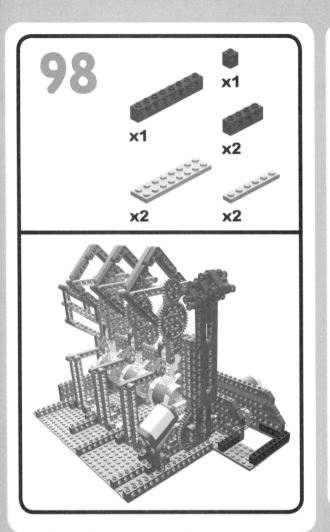

99

x6 x2

x1

x1

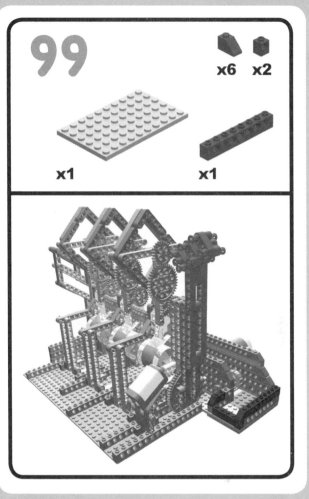

100

x1 x2 x2

101

x6 x1 x1

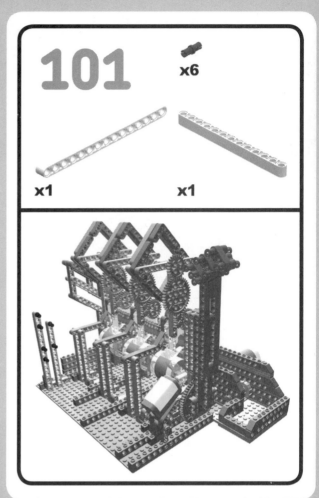

102

x8

x2   x4

x2

x2

x1

x1

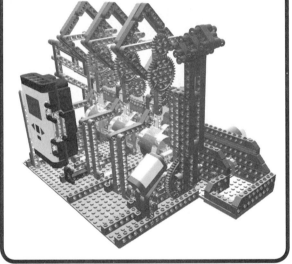

103

x1

x1

x1

x1

x2

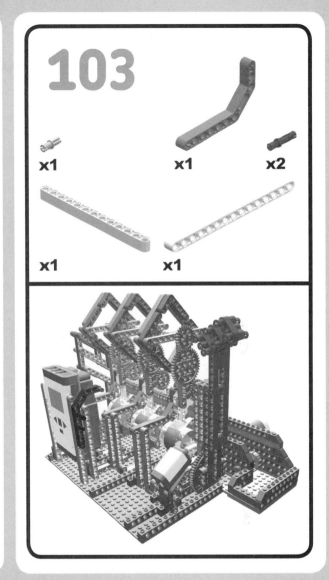

104

x2 x2

x1 x1

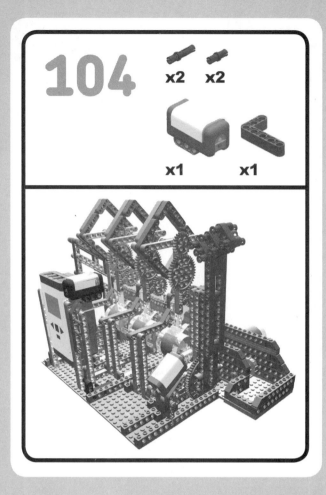

105

x6

x1 x1

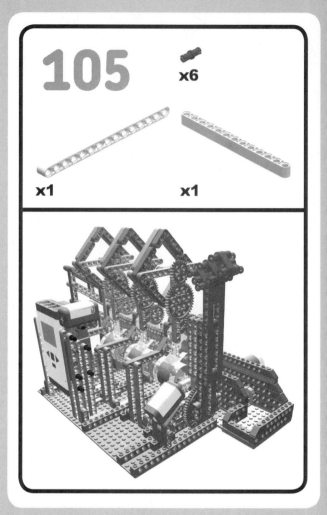

106

x2 x5

x4 x1

x2 x1

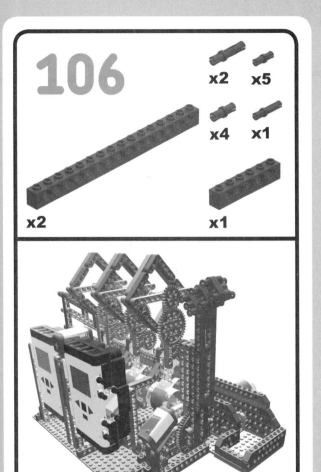

107

x2

x1 x2

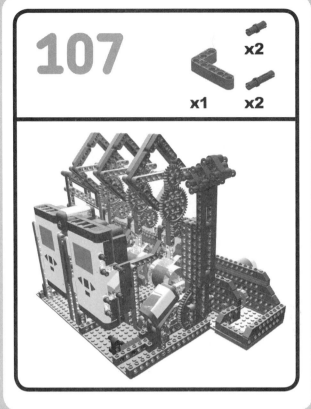

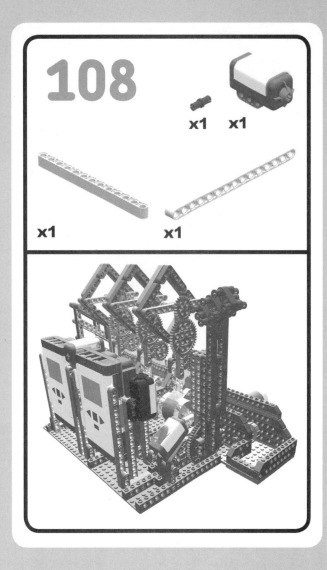

finishing up

Now that you've completed the building steps for this robot, we have only two things left to do: add symbols and connect the motors and sensors.

adding symbols

We have to put some symbols on the four sides of each reel. Choose whichever symbols you like (for instance, the stickers in the NXT kit); just make sure you use the same set of symbols for each reel and attach each symbol in the same order around the reel.

connecting the motors and sensors

Now we need to connect the motors and sensors to the two NXT bricks. Follow these five steps and see Figure 13-10 for guidance.

1. Connect the motors of the reels to the output ports of the NXT brick that is installed on the left side of the Slot Machine (as viewed from the front). Use the cables to wire each reel to one of the output ports A, B, and C.

2. Connect three of the four sensors to the left NXT brick.

   * Connect the Touch Sensor that is located beneath the lever to input port 2.

   * Connect the slot's Light Sensor to input port 1.

   * Connect the Touch Sensor next to the right brick (the Cancel button) to input port 3.

3. Connect the two motors of the pay-off mechanism to the NXT brick on the right side of the Slot Machine.

4. Connect the left motor to output port B and the right motor to output port C on the right brick.

5. Connect the remaining flashing Light Sensor to input port 1 of the right brick.

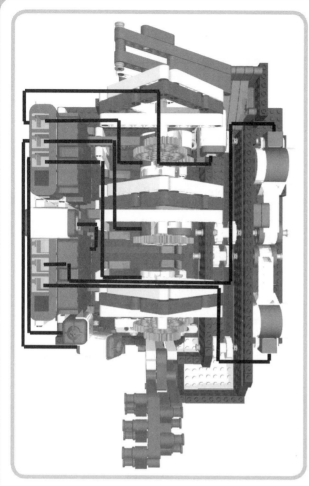

Figure 13-10: Wiring the sensors and motors

general program flow

Because the Slot Machine uses two NXT bricks, we'll require two different programs, one for each brick.

We will distribute the workflow of the machine between these two programs:

* The program running on the brick that is located on the left side (viewed from the front) is the game controller. It controls the spinning of the reels, the lever, the slot, and the Cancel button.
* The program running on the right brick is the game's result controller. It is in charge of the pay-off mechanism and the flashing light.

the game controller

The game controller waits for a pull on the lever and then runs a game. It performs the following functions:

1. Resets the reels

2. Pockets a coin

3. Starts a piece of fancy music

4. Runs the reels and generates random numbers for each reel to represent the number of rotations, then turns the reels accordingly

5. Stops the music

6. Checks the state of the reels

7. Sends a win or loss Bluetooth notification to the game's result controller running on the other brick

8. Concurrently, listens for the insertion of coins and for a press on the Cancel button by the gamer (which stops the game)

Figure 13-11 shows the program flow for the game controller.

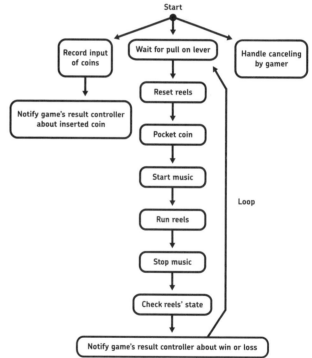

Figure 13-11: Program flow for the game controller

the game's result controller

The game's result controller is in charge of displaying wins or losses to the gamer and paying off the coins in case of a win. It performs the following functions:

1. Waits for Bluetooth notification by the game controller about a win or a loss

2. Displays the win or loss on the LCD

3. Illuminates the flashing light in the case of a win

4. Plays an appropriate sound

5. Pays off a certain number of coins in the case of a win

6. Concurrently, waits for a Bluetooth notification by the game controller about insertion of coins and updates the coin store appropriately

Figure 13-12 shows the program flow for the game's result controller.

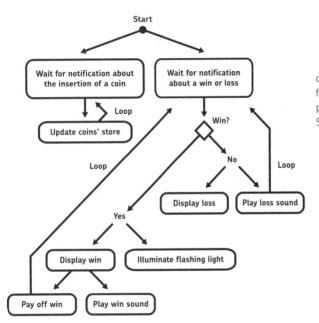

Figure 13-12: Program flow for the game's result controller

programming the slot machine

Now we will implement the two programs displayed above with NXT-G.

the game controller

We define a number of variables in our program that help us manage the actual state of the Slot Machine. They are:

limitTurnsOfReelA, limitTurnsOfReelB, and limitTurnsOfReelC   Numbers that reflect the maximum number of rotations for each reel in a single game

turnsOfReelA, turnsOfReelB, and turnsOfReelC   Helper variables that contain the actual number of rotations already performed on a reel during a single game

numberOfCoins   A variable that stores the number of coins the user has inserted minus the ones he has already used up for games

won   A logical value that is set to *true* or *false* according to the result of the actual game

To make the program easier to read and to reduce its top-level complexity, we implement a number of custom My Blocks that perform certain subtasks. We will use them to assemble the complete program. For an introduction to My Blocks, please see "My Blocks Save Time and Simplify Your Programs" on page 16.

resetting the reels

To reset the reels, we define a My Block (shown in Figure 13-13) that runs each motor backwards for the number of rotations the last run has set in the variable limitTurnsOfReel<port>. If, for instance, reel A has run for 10 rotations in the last game, the limitTurnsOfReelA will have the value 10—the block in question will run the reel A for 10 rotations backwards.

In a nutshell, this block ensures that the reels are set to the initial state.

pocketing a coin

We define a My Block that pockets a coin by decrementing the variable numberOfCoins by 1, as shown in Figure 13-14. Hence, the variable numberOfCoins contains the number of coins that are still available to the player for future games.

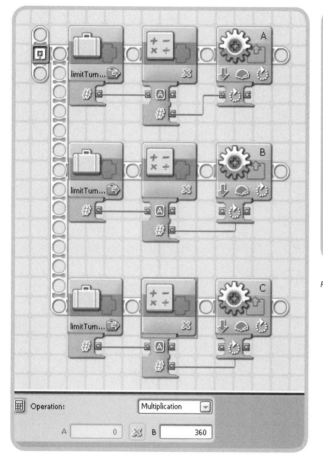

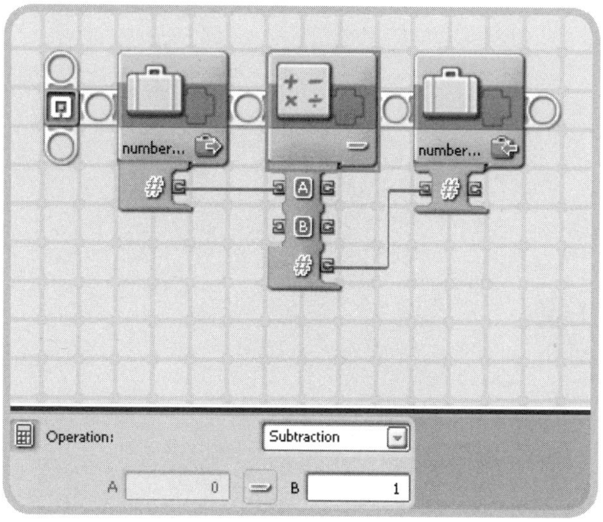

Figure 13-14: The game controller's My Block for pocketing a coin

Figure 13-13: The game controller's My Block for resetting the reels

turning the reels

We define a My Block that turns the reels. For each reel, we generate a random number between 1 and 12 (using the Random block) and run the corresponding reel for the generated number of rotations. For example, if the random number generator delivers the number 7 for reel A, that reel is rotated seven times. The runReels My Block uses a total of three runReel My Blocks, one for each reel, as shown in Figure 13-15.

It loops until the variable turnsOfReel<*port*> is greater than limitTurnsOfReel<*port*>. In each loop step we increment turnsOfReel<*port*> by one and run the reel's motor for one rotation, as shown in Figure 13-16.

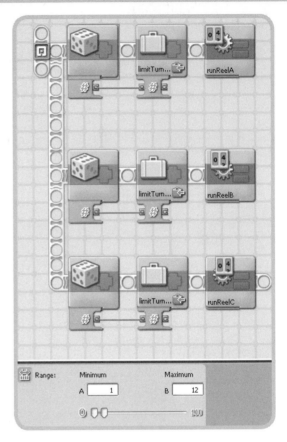

Figure 13-15: The game controller's My Block for running the three reels

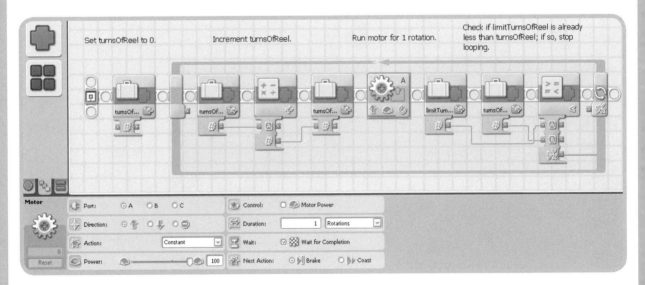

Figure 13-16: The game controller's My Block for running a reel

checking the reels' state

We define a My Block that checks whether all reels are showing the same symbols, as shown in Figure 13-17. Essentially, it checks to see if the remainders of every reel's number of rotations (contained in the actual value of the variable limitTurnsOfReel<*port*>) divided by four (the number of sides of each reel) are equal (an operation called *modulo 4* in mathematics) to each other. That is, we compare limitTurnsOfReelA mod4 with limitTurnsOfReelB mod 4. If these values are equal, we do the same for reel C—limitTurnsOfReelC mod 4. The result is stored in the logical variable won.

A little example might be of help here. Say that reel A has run 5 rotations, reel B, 9, and reel C, 11. Hence, limitTurnsOfReelA is 4, limitTurnsOfReelB is 8, and limitTurnsOfReelC is 11. Determining the remainder of the division of each of them by 4, we get 1, 1, and 3. We deduce that reels A and B show the same symbol, but not reel C. Therefore, this is not a win and, consequently, won is set to *false*.

To reduce the complexity of the block, we define a small helper My Block that computes the remainder of integer division of a number (stored in the variable dividend) by 4 and assigns it to the variable remainder.

We use that helper block in the My Block for checking the reels, as shown in Figure 13-18.

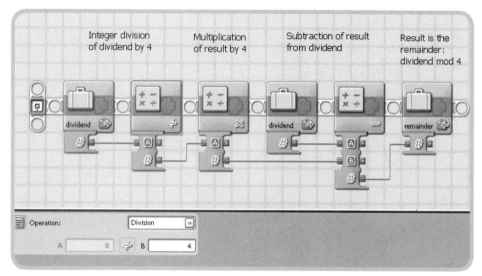

Figure 13-17: The game controller's My Block for the modulo 4 operation

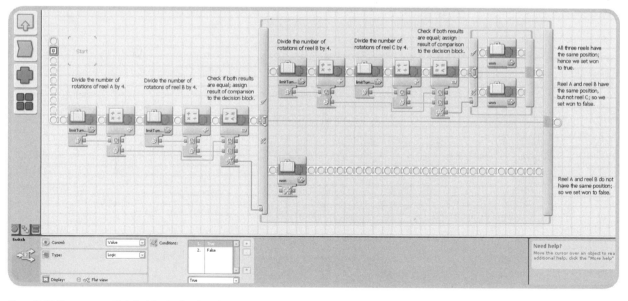

Figure 13-18: The game controller's block for checking the reels

sending a win or loss bluetooth notification

We define a My Block that sends a win or loss notification to the game's result controller running on the other brick. We assign the content of the variable won to the Logic slot on a Send Message block and send the message to mailbox #1 on connection [1], as shown in Figure 13-19.

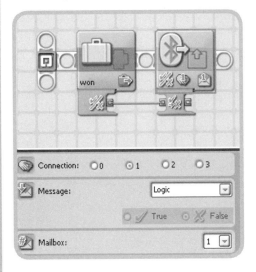

Figure 13-19: The game controller's My Block for sending a win or loss notification

recording inserted coins

We define a My Block that records and counts inserted coins, as shown in Figure 13-20. The Light Sensor attached to the slot is triggered when a passing coin reflects the light sent out by the sensor, and the variable numberOfCoins is incremented. Additionally, we send the actual number of coins to the program running on the other brick via Bluetooth (again on connection [1], but this time to mailbox #2).

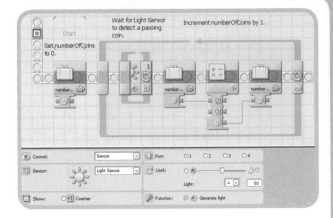

Figure 13-20: The game controller's My Block that counts the coins inserted by the player

canceling

We define a My Block that cancels the complete game when the user presses the Cancel button, as shown in Figure 13-21. The Touch Sensor on input port 3 is checked for press events. Once the user presses the sensor, a sound is played and a Stop block terminates the program.

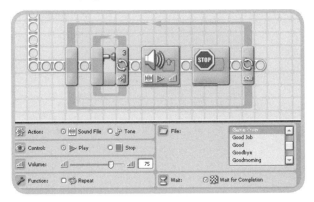

Figure 13-21: The game controller's My Block that cancels the whole game

The complete program for the game controller running on the left brick is shown in Figure 13-22.

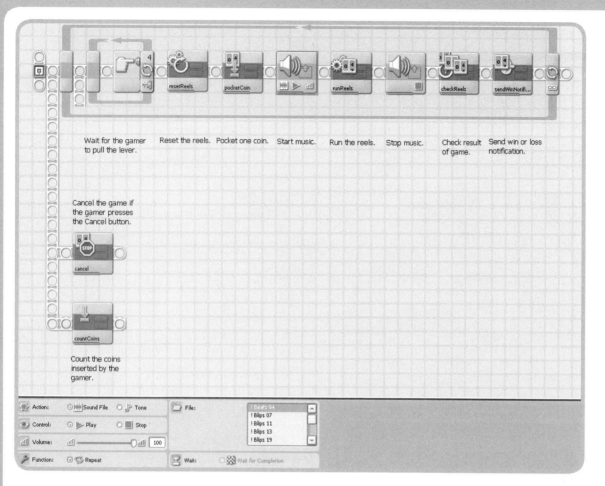

Wait for the gamer to pull the lever.

Reset the reels.   Pocket one coin.   Start music.   Run the reels.   Stop music.   Check result of game.   Send win or loss notification.

Cancel the game if the gamer presses the Cancel button.

Count the coins inserted by the gamer.

Figure 13-22: The complete program for the game controller

the game's result controller

As with the game controller, we needed some helper variables when programming the game's result controller.

numberOfTreadSteps   The number of steps the pay-off mechanism's treads are to be run; a step rotates the motors attached to the treads by 180 degrees (which moves the treads forward for approximately the size of one coin)

numberOfCoins   The number of coins actually contained in the store

positionOfLeadingCoin   The position of the leading coin in the store (starts with 1, if there are any coins)

the flashing light

We define a My Block that runs the flashing light by switching the Light Sensor frequently off and on for 1/8 of a second each, as shown in Figure 13-23.

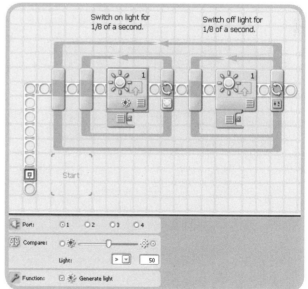

Switch on light for 1/8 of a second.   Switch off light for 1/8 of a second.

Start

Figure 13-23: The game result's controller's My Block that flashes the light

the pay-off mechanism

To manage the pay-off mechanism, we first define a My Block that runs the treads: The two attached motors are run for the number of rotations that is set in the variable numberOfTreadSteps, as shown in Figure 13-24. (Note that the two motors have to run in opposite directions due to the way they are attached to the two treads.)

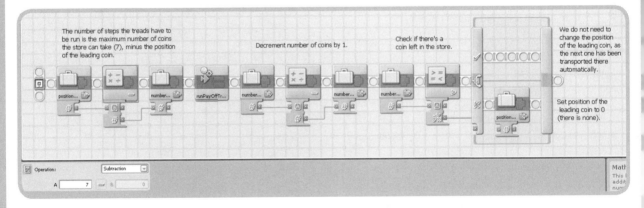

Figure 13-24: The game result's controller's My Block that runs the pay-off treads

Next, we need a My Block that pays off one single coin. First, we have to run the treads for the number of steps it takes to transport the leading coin in the store to the output slot. Next, we decrement the variable numberOfCoins by 1 (we have paid off one just now) and adjust the variable positionOfLeadingCoin, if required, as shown in Figure 13-25.

Finally, we define a My Block that in the case of a win plays a pleasant sound, illuminates the flashing light, and pays off the win (three coins), as shown in Figure 13-26.

Figure 13-26: The game result's controller's My Block that handles a win

updating the store

The store has to be updated once the gamer inserts a coin; that is, when a Bluetooth notification arrives from the game controller running on the other brick. When a coin is inserted, we have to transport it for one step on the treads to provide an empty space in the store, should another coin arrive (see Figure 13-27). Note that if the store is already full, the leading coin will be disbursed automatically by falling from the tread into the disbursement device. We have to consider that case when adjusting the variable numberOfCoins.

Figure 13-25: The game result's controller's My Block that pays off one single coin

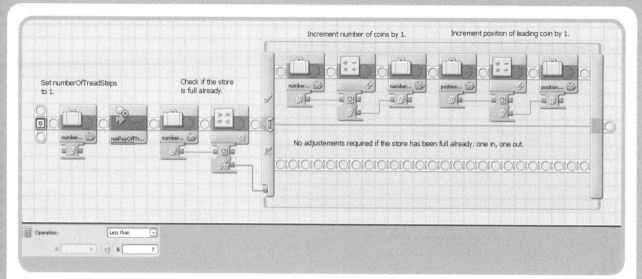

Figure 13-27: The game result's controller's My Block that updates the coin store

Finally, we assemble the complete program for the game result's controller running on the right brick, as shown in Figure 13-28.

running the programs

Before running the programs, make sure that you have connected the NXT brick that is located on the left side (the game controller) to the other brick (the game result's controller) via Bluetooth and that the reels are pre-set to show equal symbols.

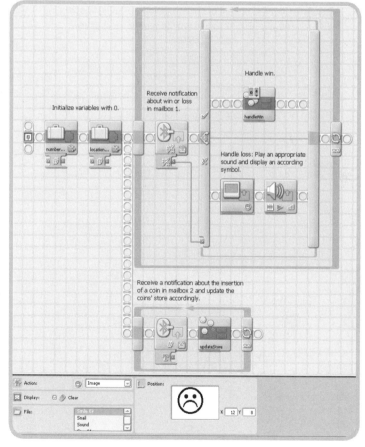

Figure 13-28: The complete program for the game result's controller

ideas for enhancing your robot

Now that you have built, programmed, and, I hope, had fun with the Slot Machine, I imagine you have noticed that there are various ways to enhance it.

Here are some ways to do a bit more with your creation. Unlike real-world slot machines, the internals of our Slot Machine are not hidden from the player's view. We have left them exposed because we think it's more interesting for a user to see how the robot actually works. Still, you might choose to make the Slot Machine look more like the real thing by covering it with bricks, plates, and beams, allowing the user to only see the reels' front pages.

Next, you might design an additional component for the coin's store to separate it from the transportation treads. This would allow the Slot Machine to pocket, hold, and pay off a much greater number of coins.

As with the slot machines you find in casinos, our model might be enhanced with fancy sounds, lights, and displays to be triggered by either a win or loss. These could greatly enhance the gaming experience.

Last but not least, our design determines the result of a single run of the game based on *dead reckoning*: Our robot decides whether the reels show the same symbol by computing the internally stored number of turns of each reel. But dead reckoning is not the best of all strategies in robotics; small differences in the performance of the motors or unforeseen hardware issues may result in a final state of the reels that does not reflect the computed result. As a consequence, the Slot Machine may show a win even if the reels do not, or (worse for the gamer), the reverse! A more reliable method of confirming the state of the reels would be desirable, such as a set of hubs on each reel that a Light Sensor could check.

And let's not forget security.

Our Slot Machine is completely lacking in security. We don't check to see if the gamer actually inserted enough coins to play the present game or to receive the three coins of a win. We should certainly consider security aspects in a future version of the program.

And how about a more sophisticated win table? With our current design, a win always spawns the same pay-off—three coins. Yet, contemporary slot machines provide much more complex strategies for betting, winning, or losing.

Certain constraints must be observed before running the current machine: The reels have to be set to show equal symbols at the beginning, and the coin store must be empty. Future versions of the program might provide features that could deal with reels in different positions and coins already existing in the store.

As you can see, many improvements could be made to the Slot Machine presented in this chapter.

If you are so inclined to improve this robot on your own, please do so, and share your results and ideas with us!

BenderBot: an anti-theory music robot

American composer and philosopher John M. Cage said, "There is no noise, only sound." From widely accepted traditional and contemporary to controversial chance and electronic, infinitely variable forms of music exist in the world today. With this bit of wisdom in mind, it is possible to create music with your LEGO MINDSTORMS NXT kit.

Thinking outside of the box with what's inside the box, this chapter is about building and programming a new nonstandard musical instrument: the experimental sound generation machine, BenderBot (see Figure 14-1). Additionally, you will be exploring and learning musical anti-theory while using the NXT motors and sensors to bend sounds and produce some wild music called *Bent Music*.

Now let's create our instrument.

Figure 14-1: BenderBot

DISCOVERING BENT MUSIC

Music theory is a widely accepted field of study that investigates the nature or mechanics of music. An extensive technical language is used to describe and explain music, often involving identifying patterns that govern composers' techniques. In a more general sense, music theory also often distills and analyzes the elements of music: rhythm, harmony, melody, structure, and texture. However, some would argue that it can also inhibit creation and the exploration of new music genres.

Bent Music represents musical anti-theory. Derived from *circuit-bending*—the creative art of short-circuiting low voltage, battery-powered electronic audio devices (such as guitar effect pedals, children's toys like the Texas Instruments Speak & Spell, and small electronic synthesizers) to create new musical instruments and sound generators—Bent Music breaks the traditional interpretation of what is usually considered music.

building BenderBot

The BenderBot configuration illustrated in the following building instructions is designed for a right-handed musician. If you're left-handed, you can easily alter the configuration by flipping the sensor mounts and the center motor any way you like.

There are many ways to build your own BenderBot. The sensors, motors, and NXT brick itself offer many arrangement possibilities; and you might find that altering the configuration to fit your "playing" style will ease your music production efforts. I recommend starting with the following BenderBot configuration and then experimenting with some of the possibilities that follow.

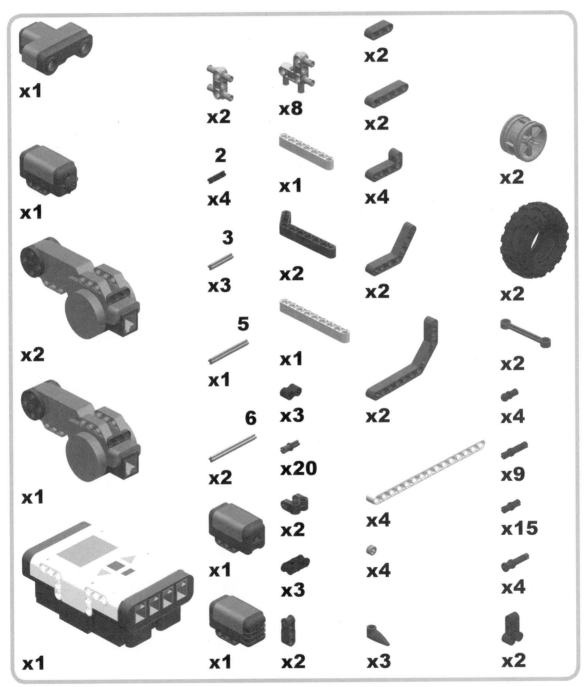

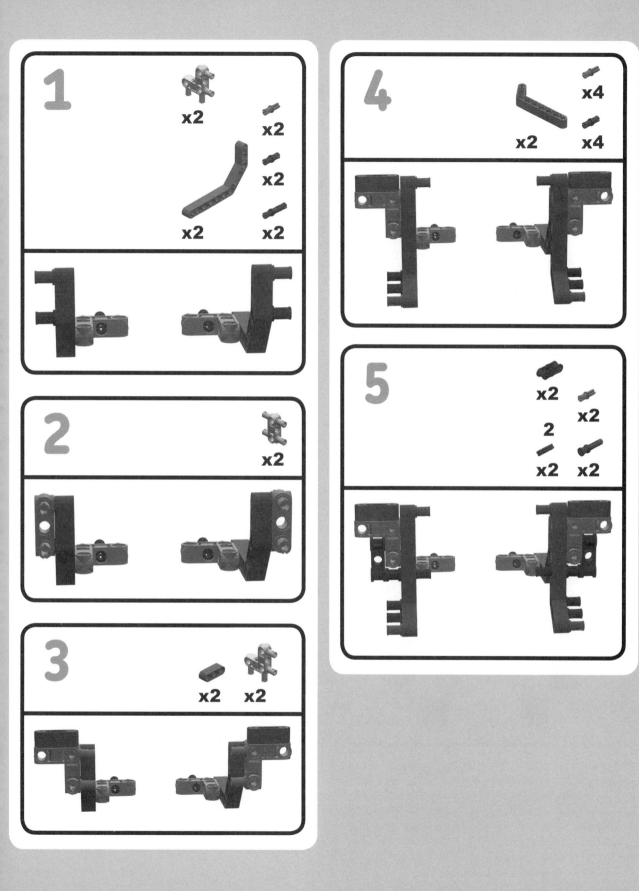

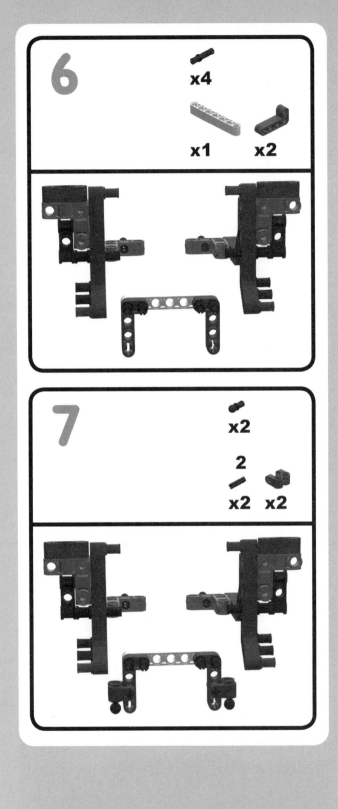

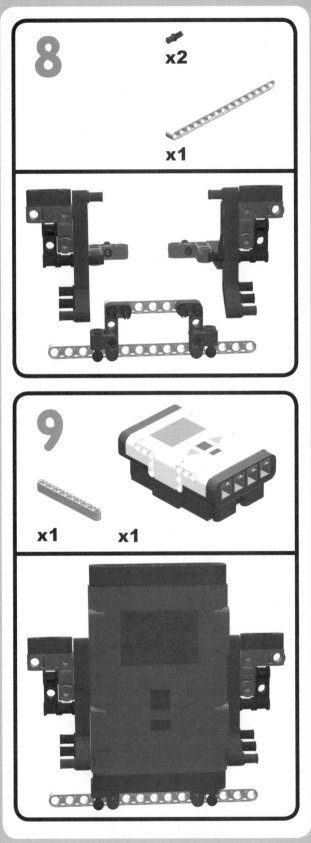

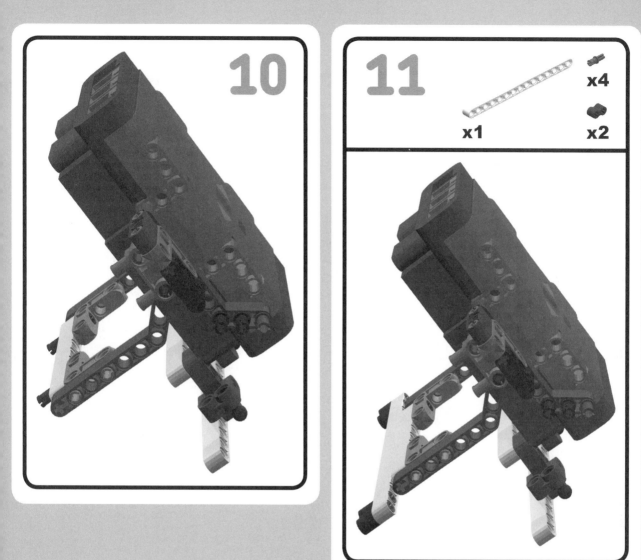

12

x2

3
x1 x2 x1

x4
x2
x1

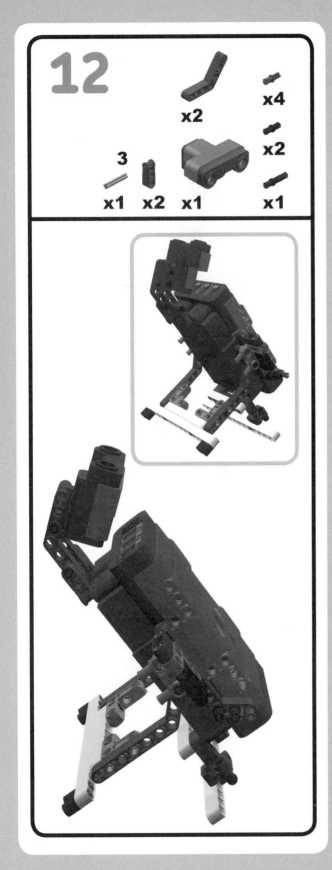

13

x1 x1

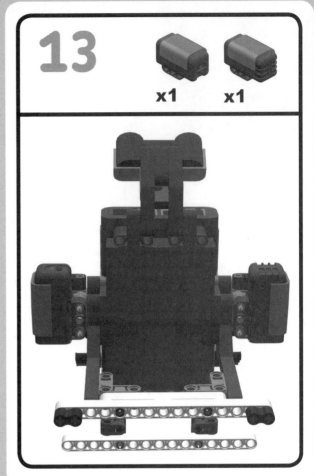

14

x4 x2

x1 x2

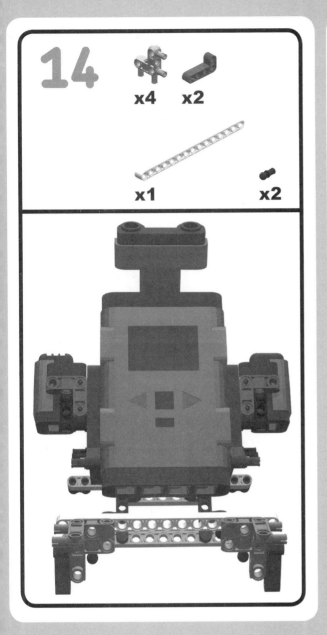

15

x2

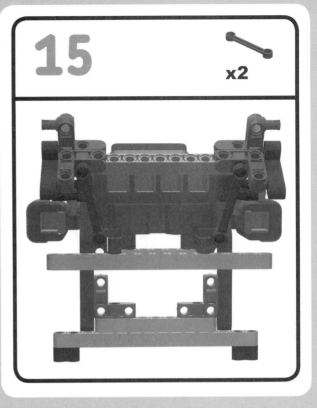

16

x1
6 x1
x1 x1 x1
x2
x1

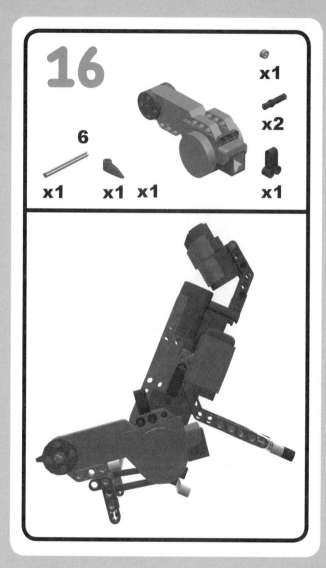

17

x1
x1 x1

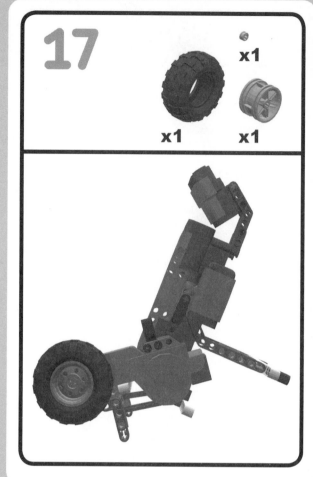

18

x1

6

x1 x1 x1

x1

x2

x1

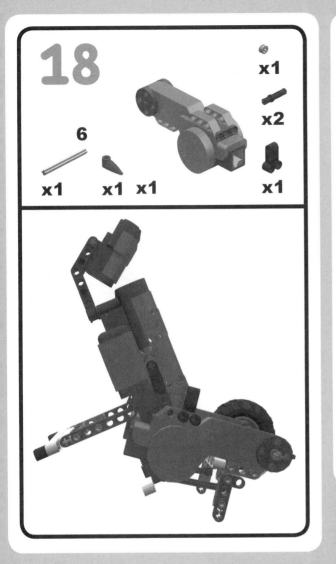

19

x1

x1 x1

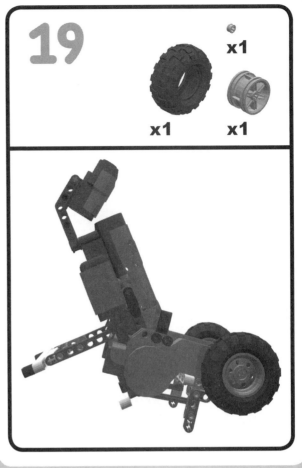

20

x2

x1

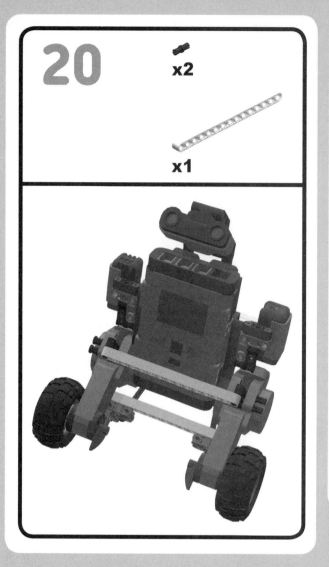

21

5   3

x1   x2   x1

x3

x1

wiring BenderBot

Finally, grab your wiring and complete BenderBot by attaching the sensors and motors to the standard NXT brick ports, as described on pages 7 and 8 of the LEGO MINDSTORMS NXT User Guide. After you have routed the wiring *cleanly*, you are ready to program BenderBot.

programming BenderBot

Here you will learn a few sound-generating techniques that use the standard LEGO NXT-G programming software. This section is not limited to the sound production solutions described here. After you learn what can be produced with the following examples, you should further explore the limitless possibilities. While programming BenderBot you will find that anything can happen to fit your musical style better.

Before you can program your BenderBot, you will need to install the Dynamic Block Update and then the Mini Block Library

found at http://mindstorms.lego.com/support/updates. These updates will save valuable NXT memory space and allow more room for large sound files and complex musical programming.

a program to start you off

The BenderBot program illustrated in Figure 14-2 presents a combination of several sound-generating code examples. Re-create the code with NXT-G, upload it to your NXT, and then run it.

As you'll see, the program produces a constant backbeat of three different drumbeats that play randomly. Begin to add effects by waving your hand above the Ultrasonic Sensor that sits atop BenderBot. This action, which resembles the function of a theremin, will produce varying sound frequencies and pitches determined by the distance between your hand and the sensor. (A true *theremin* is a musical instrument designed to be played without being touched and consists of two radio frequency oscillators and two metal antennae.)

The Magic Light is also included for sound production. Detach the Light Sensor from BenderBot and hold it over the colors on the large NXT kit's fold-out test pad poster or even the reflective hologram security patch of a credit card to produce some very exciting sounds. (You will need to press and hold the Touch Sensor's button when using the Magic Light.)

This program is one example of exploring the vast musical possibilities you can generate with the LEGO MINDSTORMS NXT kit. Listed as follows are some more experiments awaiting exploration.

CREATE YOUR OWN CUSTOM SOUND FILES

You can create your own custom sound files by converting any WAV sound file into NXT-G RSO sound files with Wav2rso (a freeware program you can download from the Internet). Every PC has a sound recorder (Windows machines contain one that records sounds from a microphone and saves them as WAV files). Grab your microphone and create some new sounds for BenderBot.

After you have created a new RSO sound file, save it to the default NXT-G sound folder located at C:\Program Files\ LEGO Software\LEGO MINDSTORMS NXT\engine\Sounds (for Windows installs). When you open the NXT-G software you should be able to choose your own sound for use within your own Sound blocks. Try experimenting with different sound files and see what sorts of music you can create.

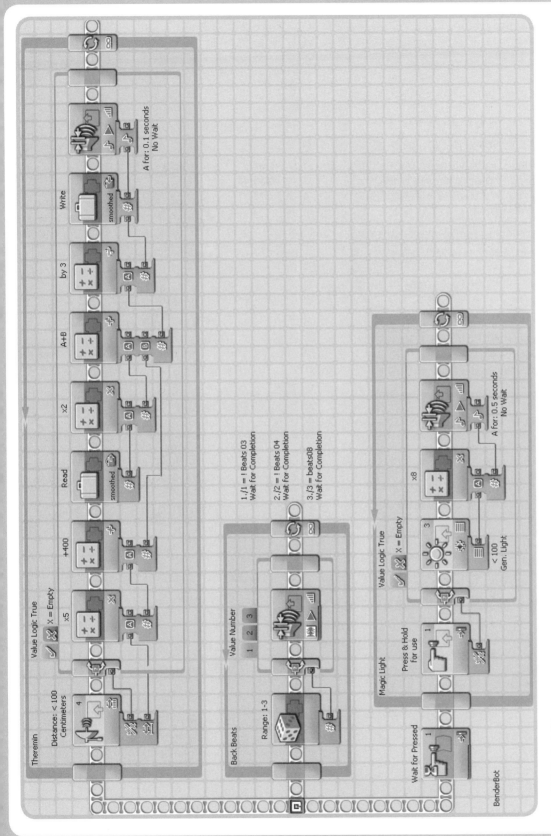

Figure 14-2: The BenderBot program

additional programs to explore

You can produce some very interesting sounds and effects with the NXT-G software. A few more examples for interesting sound production are illustrated in this section. Create, combine, and explore using these examples. And remember: Music is what you make of it.

BeatBox program

For example, the BeatBox program illustrated in Figure 14-3 is a very interesting one that produces unique sounds that are extremely fun to play with. With your hands, manipulate the sounds by using the motors attached to the NXT brick's A and C ports to produce a looping sound and adjust the sound length and delay period of the tone being produced. BenderBot doesn't roll with its wheels . . . it rocks!

Give other sounds a try by loading them into that first Sound block. You can use any of the NXT-G sounds in the sound library that were loaded when you installed the LEGO software. I recommend starting with the sounds *Backwards*, *Error02*, *!Sonar*, and *LEGO*, but you are not limited to these sounds. As you have learned, you can use Wav2rso to create your own NXT-G sound files for BenderBot.

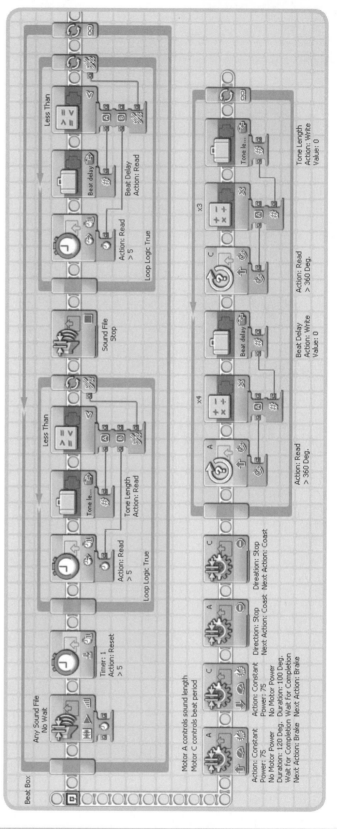

Figure 14-3: The BeatBox program

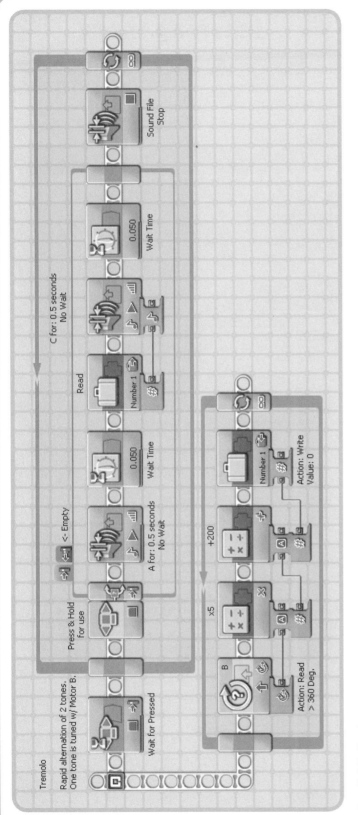

Figure 14-4: The Tremolo program

tremolo program

The Tremolo program illustrated in Figure 14-4 resembles the BeatBox program in that a motor is used to manipulate sound. Two rapidly alternating tones are produced, one of which can be altered using motor B. Re-create this code to discover how your NXT can produce tremulous effects.

chance program

The Chance program illustrated in Figure 14-5 explores an interesting musical genre called *aleatoric music*. Sometimes called *chance music*, this is music in which some element of the composition is left to chance. Re-create this code and explore some nonstandard music theory.

basic theremin program

The Basic Theremin program illustrated in Figure 14-6 is much like the theremin used previously because the concept is the same. This example is simpler and produces a different sound. Re-create this code and discover the difference.

learning more

Visit this book's companion website for additional sound files, programs, discussions, and fun for your BenderBot! It is a gathering place to share your musical creations as well.

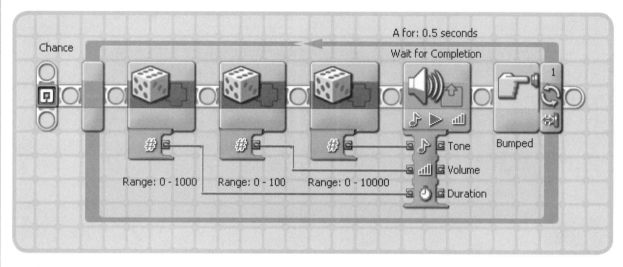

Figure 14-5: The Chance program

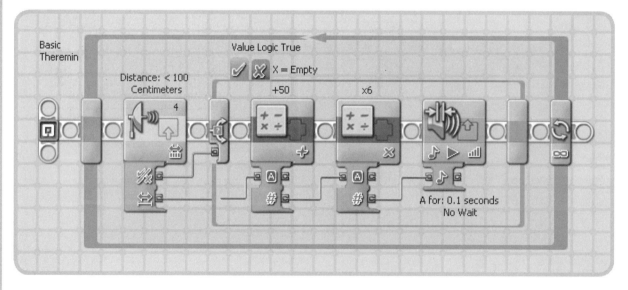

Figure 14-6: The Basic Theremin program

15

ScanBot: an image-scanning robot

Before the NXT was released, I was always impressed when someone made a robotic scanner. Since scanners are used in everyday life and require very precise measurements to function properly, they're some of my favorite robots. But since the RCX in the RIS set had very limited display capabilities, designers of LEGO scanners had to come up with a way to display the scanner's picture on a computer monitor instead of on the robot itself.

When the NXT came out with enhanced display capabilities, I immediately saw a great opportunity to build a simple robotic scanner. After a few tries, I succeeded with ScanBot, a robot that can scan black-and-white pictures and display them right on the NXT's LCD.

This chapter will show you how to build and program your own ScanBot (see Figure 15-1), using only one basic MINDSTORMS NXT kit.

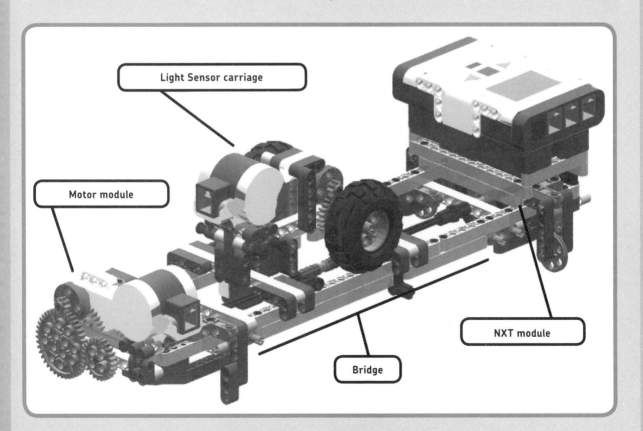

Figure 15-1: ScanBot

building ScanBot

ScanBot has a simple design and is quite easy to build. It works by moving a Light Sensor across every part of an image to be scanned, taking hundreds of measurements as it goes. It compares each measurement with a variable threshold and determines whether to represent the tiny area from which the measurement was taken with a black or white dot on the NXT's LCD. A unique and fun aspect of ScanBot's design is that the entire robot moves over the image to be scanned, instead of feeding the image through the robot or moving the Light Sensor over the image within the robot.

ScanBot has four main components, which are labeled in Figure 15-1. In the middle of ScanBot, two long beam constructions form the bridge, which supports the Light Sensor carriage, a motorized carriage that holds a Light Sensor and travels back and forth across the bridge to scan successive lines of the image. Each end of the bridge is supported by a wheel: one under the NXT module, and one (which isn't visible in Figure 15-1) under the motor module. The two wheels are connected by a drive shaft, which the motor module rotates to move the entire ScanBot across the image. To scan, the Light Sensor carriage travels over the bridge, scanning one line of the image. Then the motor module moves ScanBot down a tiny bit, and the Light Sensor carriage travels across the bridge again, scanning another line of the image right below the previous line. This process is repeated many times, until as much of the image that can fit on the LCD has been scanned.

First, build the motor module, which moves ScanBot down the image being scanned. The motor module includes the wheel beneath it and part of the drive shaft that extends toward the other side of the bridge. This will later be connected to the wheel, which sits under the NXT module on the other side of the bridge. Both wheels will then turn the same amount, giving a preciseness to the downward movements that is necessary to get a good scan.

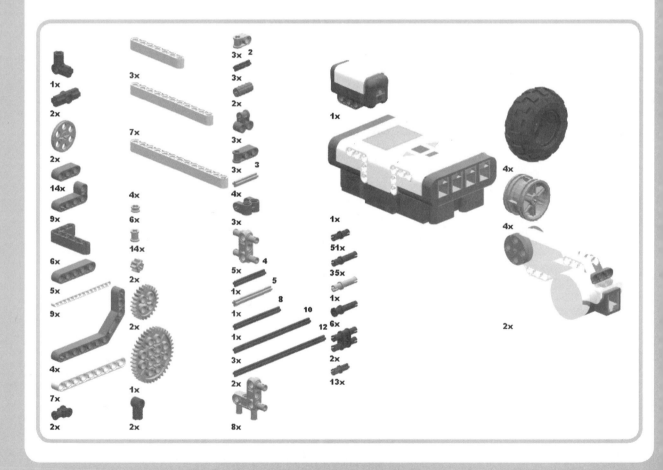

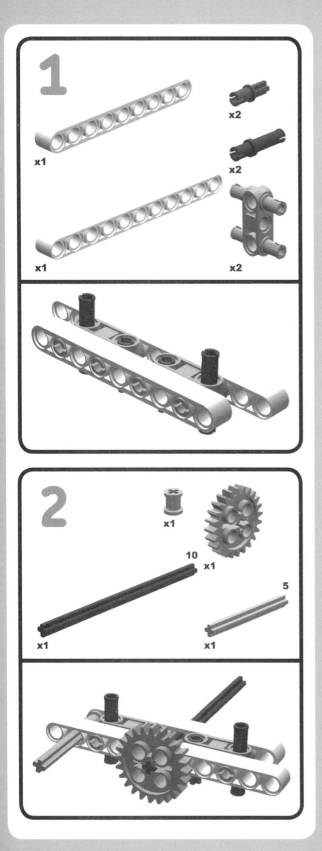

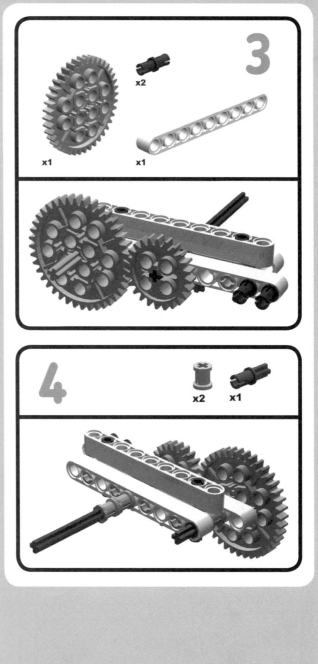

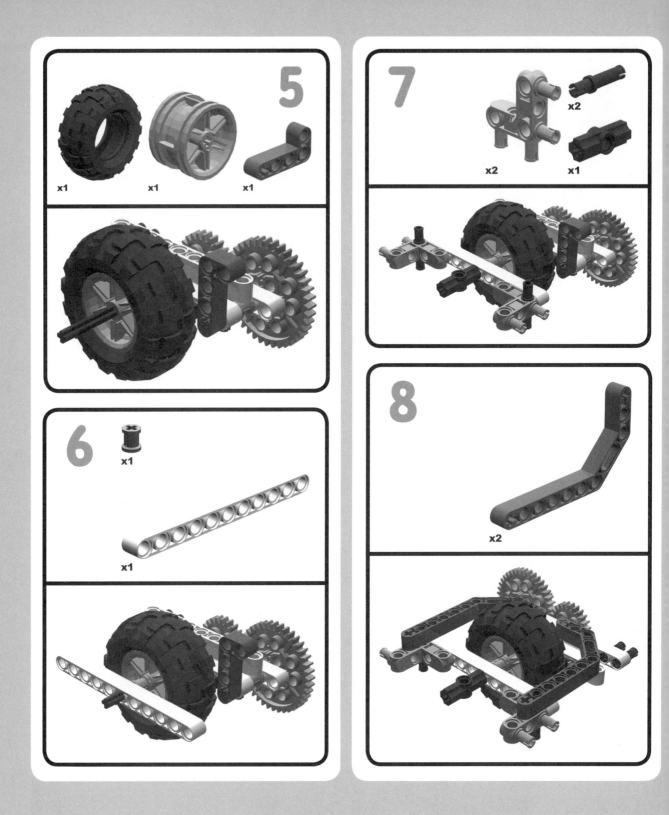

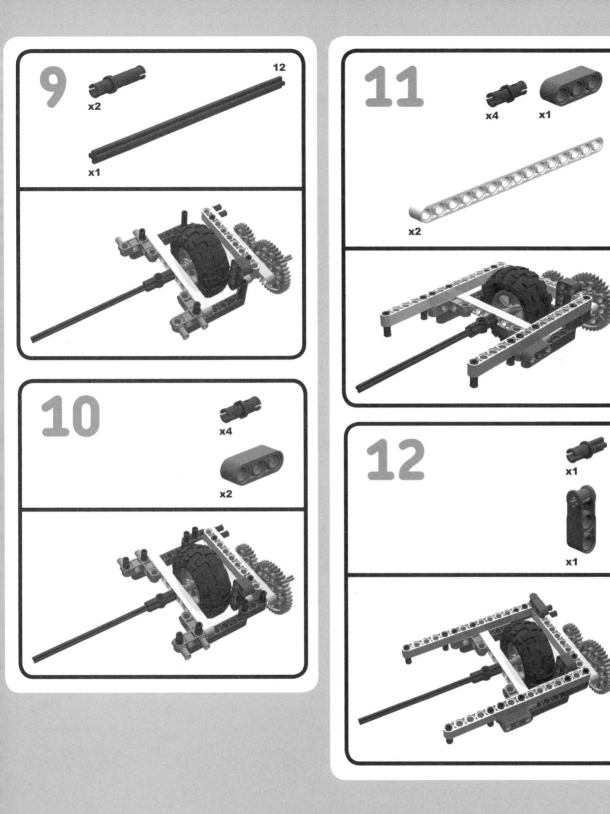

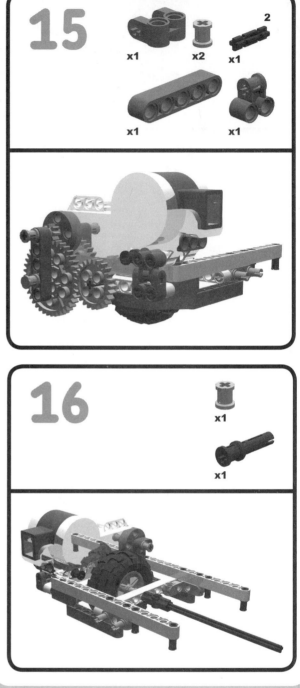

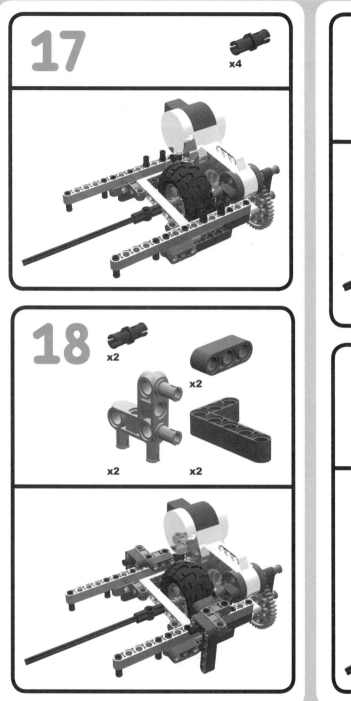

NXT module

Now build the NXT module that supports the other side of the bridge. This module contains the wheel that's driven by the motor module's drive shaft and also holds the NXT brick.

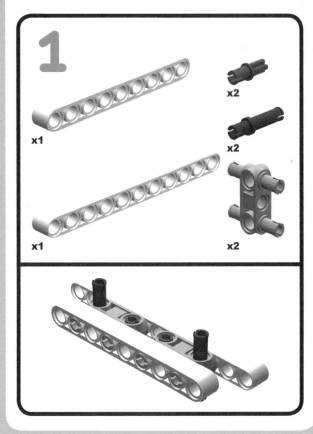

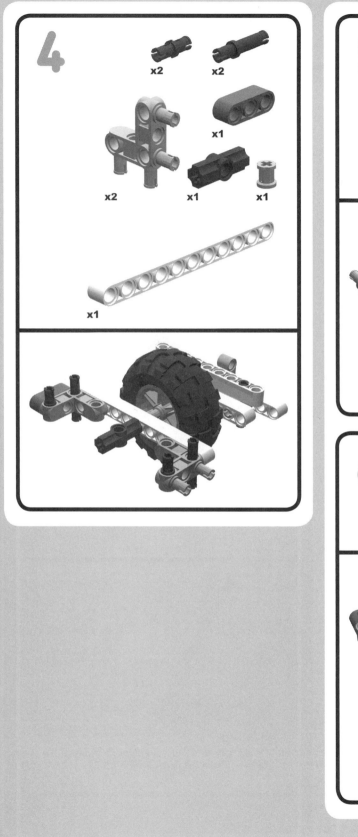

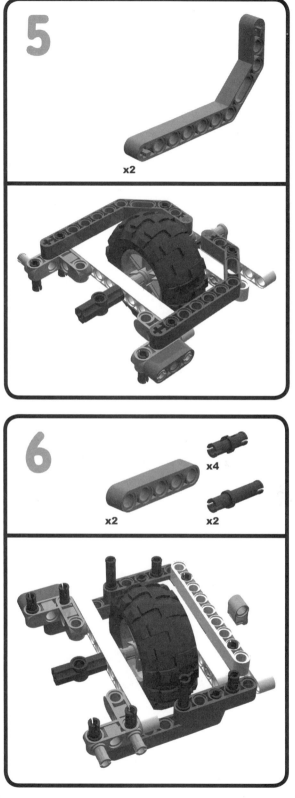

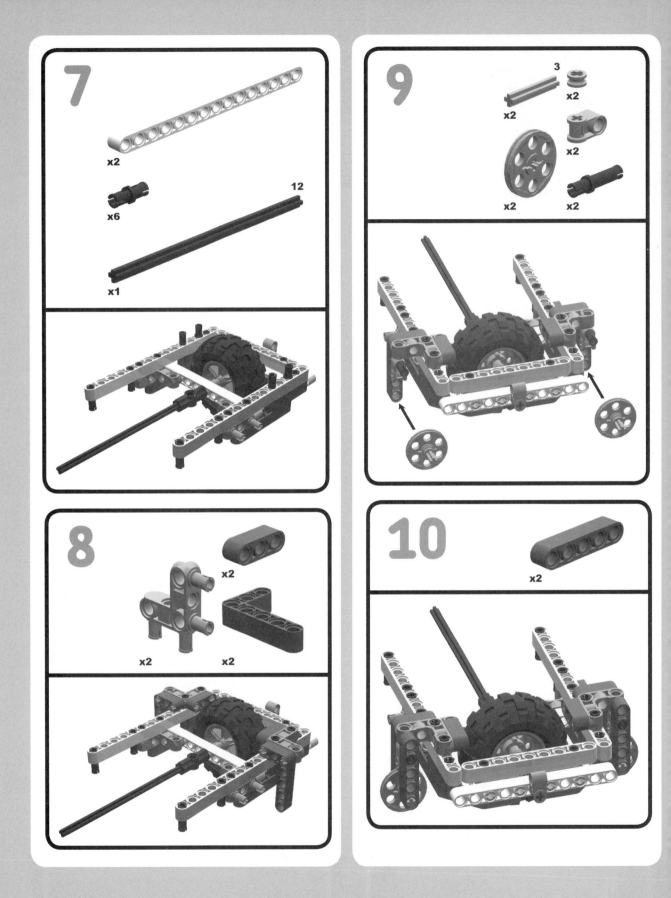

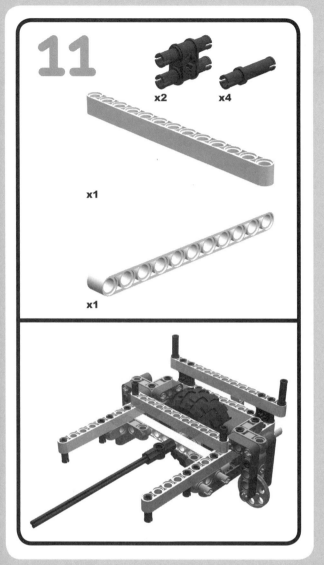

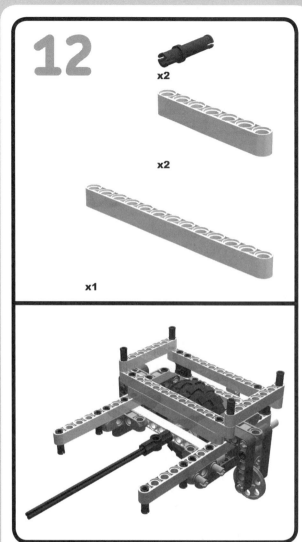

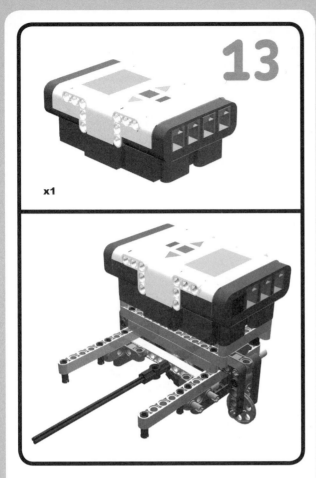

x1

x1

light sensor carriage

Now build the Light Sensor carriage, which holds the Light Sensor and travels back and forth across the bridge. Since it needs to move slowly and precisely, the gear train has a ratio of 1:3, which makes the wheels turn slowly but powerfully.

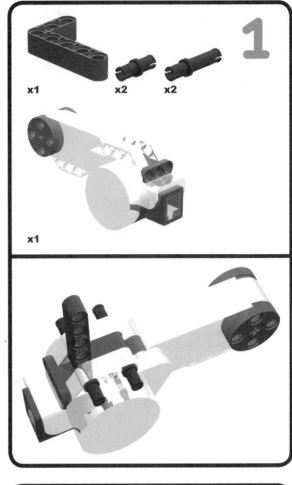

1

x1   x2   x2

x1

2

x1   x1

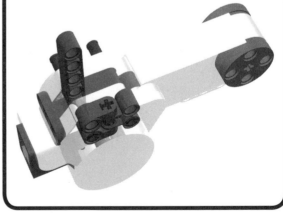

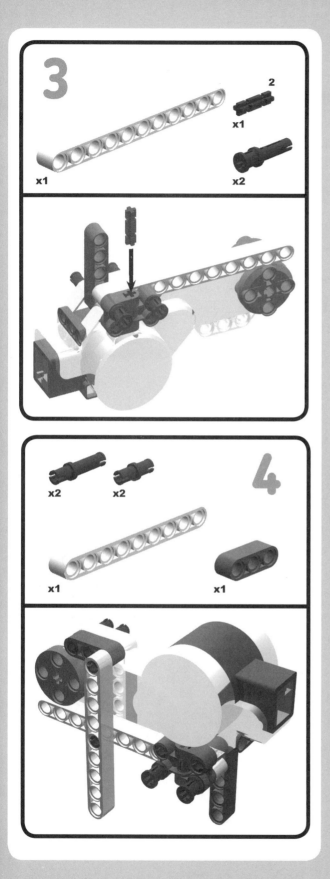

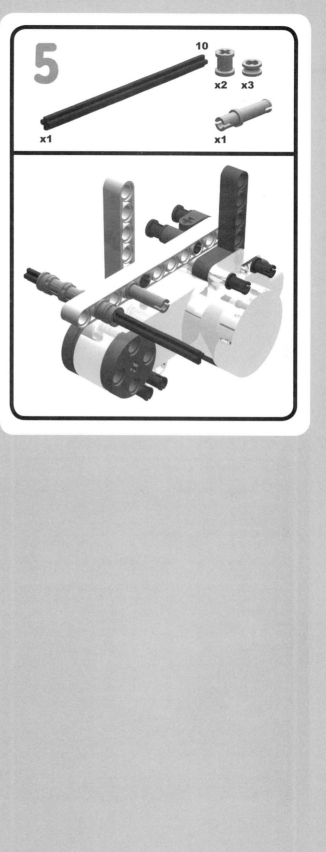

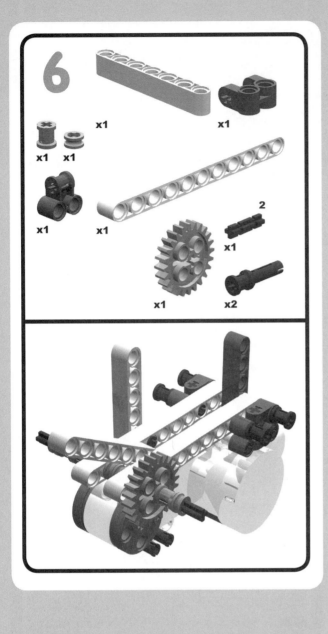

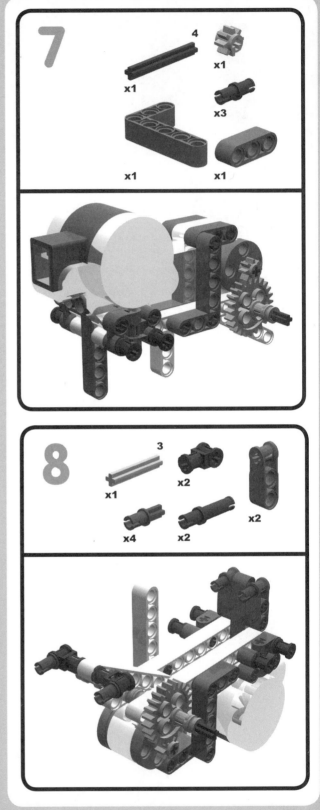

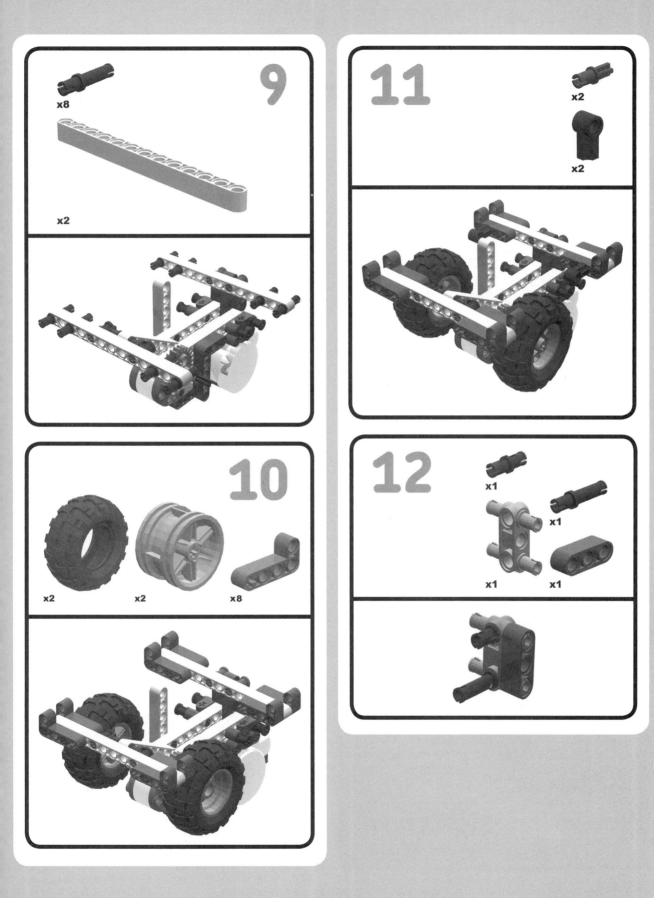

13

x1

x1

14

x1

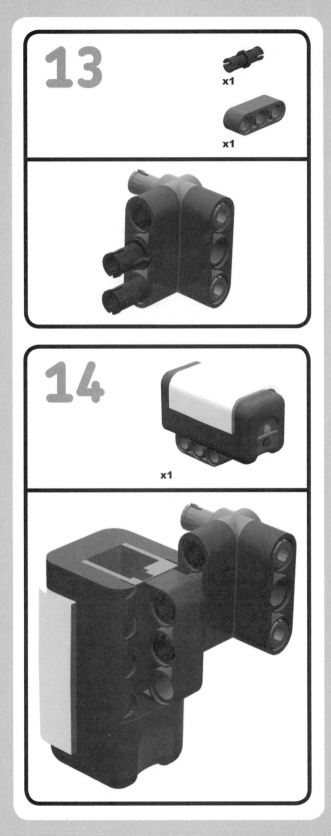

15

x1

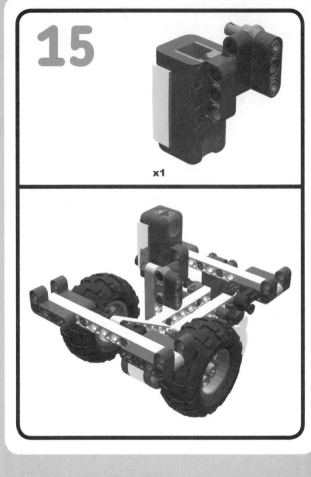

final model

Now you're ready to finish constructing ScanBot. First, you will connect the motor module and NXT module with two long beam constructions that form the bridge. Then hook up the cables, and you'll be ready to program!

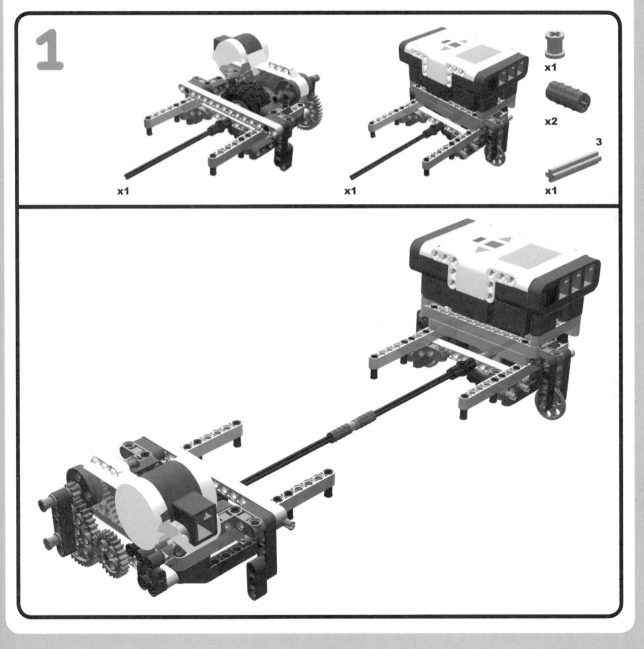

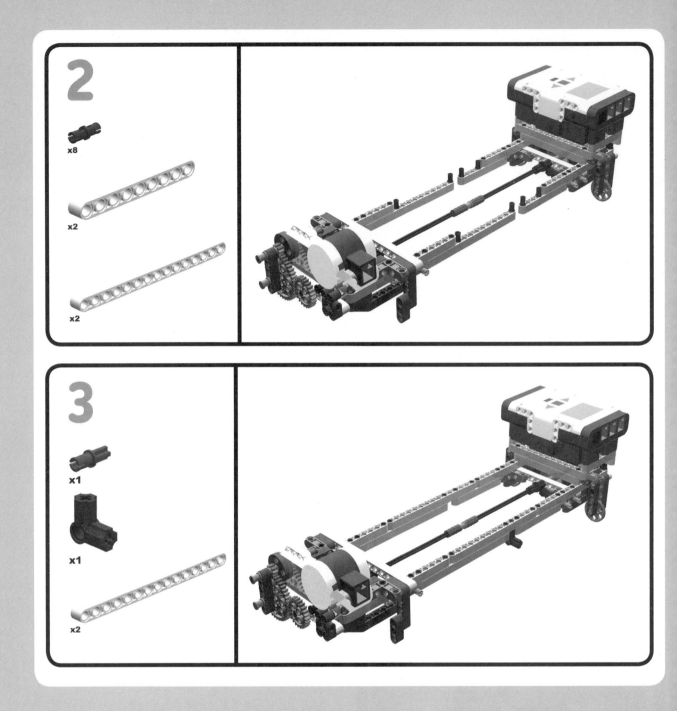

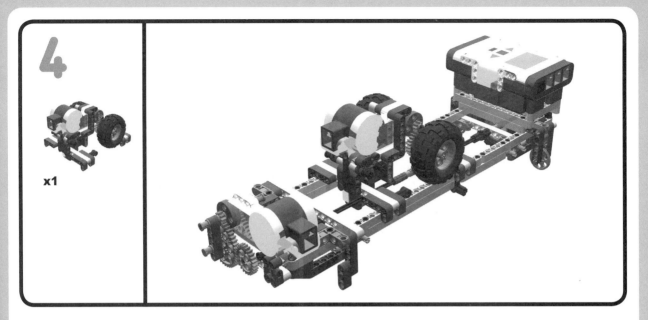

4

x1

Having completed the preceding steps, it's time to insert the cables:

1. First, insert one end of a long wire (20 inch/50 cm) into the motor that moves the Light Sensor across ScanBot, and insert the other end into output port C.

2. Insert one end of another long wire into the motor that moves the entire robot and the other end into output port B.

3. Finally, insert one end of a medium-long wire (15 inch/35 cm) into the Light Sensor and the other end into input port 3.

Make sure none of the wires rub against the wheels or prevent the Light Sensor carriage (the module that holds the Light Sensor and moves laterally) from moving across ScanBot.

Congratulations! You've finished constructing ScanBot. Now it's time to program it.

programming ScanBot

ScanBot's program is somewhat large because, besides needing to control the actual scanning of black-and-white images, it also needs to enable a user to adjust the *light threshold* (image contrast) and select a low- or high-resolution scan. Since high-resolution images take a long time to scan, the option to select a low-resolution scan can give you a rough idea of what the scanned image will look like.

defining the variables

We use several variables in our program. Before we can use them, however, we need to define their names and types.

All variables in computer programs have a unique name that is used to distinguish them from other variables. In NXT-G, they also have a type that can be Number, Logic, or Text. A variable of type *Number* stores numerical values (such as 1, 17, and −2). Variables of type *Logic* can store a logical value (*true* or *false*). Finally, variables of type *Text* can store text.

To define your variables, choose **Edit ▸ Define Variables**; then click **Create** to make a new variable. You can now enter the variables' names and types. Create the variables listed in Table 15-1, using the appropriate name and type. (I'll explain what each of these variables is used for in later sections.)

table 15-1: the variables used in ScanBot's program

variable name	variable type
Light Threshold	Number
Height Res	Number
X Position	Number
Y Position	Number
Motor Duration	Number

the my blocks

Several groups of the program blocks are in My Blocks to organize the program and show which function a specific block handles. (For an introduction to My Blocks, see "My Blocks Save Time and Simplify Your Programs" on page 16.) Let's build them.

the resolution switch my block: control ScanBot's resolution

The Resolution Switch My Block enables the user to select a low- or high-resolution scan via the NXT buttons. We'll want this My Block to offer options to choose between the Preview and High-Res options. To do this, we make the My Block detect when the Left or Right NXT buttons are pressed and then use that information to tell the main program what level of resolution to use for scanning.

1. Create a new document.

2. First, we'll program the Preview option. To do so, place a Display block configured as shown in Figure 15-2.

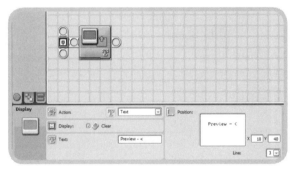

Figure 15-2: The Display block for the Preview option

3. To select a low- (Preview) or high-resolution (High-Res) scan, we use the Left and Right NXT buttons. We use Display blocks to show the user which button corresponds to each option. Figure 15-3 shows the Display block for the High-Res option.

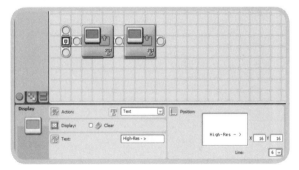

Figure 15-3: The Display block for the High-Res option

4. Add a Loop block to make the program wait until one of the NXT buttons has been pressed, as shown in Figure 15-4.

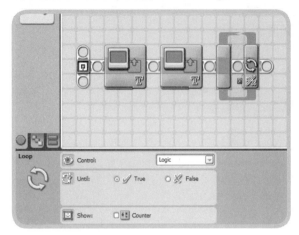

Figure 15-4: The Loop block for waiting until a button is pressed

5. Place a Sensor block configured to detect the Left NXT button being pressed, as shown in Figure 15-5.

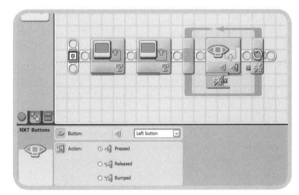

Figure 15-5: The Sensor block for detecting the Left NXT button

NOTE Don't worry that the figure shows only the logic output of the Sensor block—I just did it that way to make it look simpler. You can make your blocks look like this (it makes your programs look much tidier) by retracting a block's data hub after drawing wires from the outputs being used. All outputs and inputs that don't have wires in them will disappear, but the ones with wires will stay!

6. Place a Sensor block configured to detect the Right NXT button being pressed, as shown in Figure 15-6.

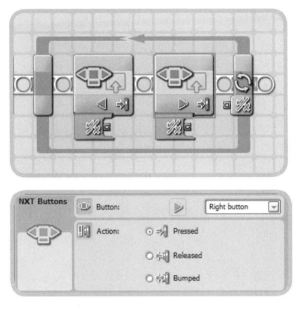

Figure 15-6: The Sensor block for detecting the Right NXT button

7. Since we want to leave the loop and continue with the program after either button is pressed, and because we can't do this with a regular Wait block, we'll add a Logic block (as shown in Figure 15-7), which performs an Or logical test. Once wired, the Logic block will output a *true* signal when either the Left or Right button is pressed and then end the loop.

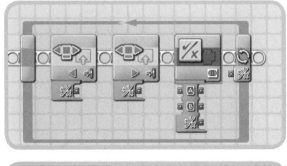

Figure 15-7: The Logic block configured to perform an Or logical test

8. Wire the Logic and two Sensor blocks (as shown in Figure 15-8) so that the Loop will end when either of the NXT buttons is pressed.

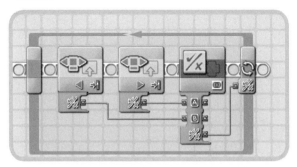

Figure 15-8: The Logic block wired with the two Sensor blocks

9. Place a Switch block configured to determine when the Left NXT button has been pressed, as shown in Figure 15-9.

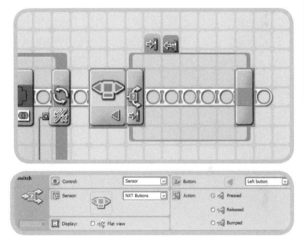

Figure 15-9: The Switch block for determining whether the Left button was pressed

10. If the Left NXT button is pressed, we want to set the resolution to low. We do this by adding a Variable block (shown in Figure 15-10) configured to set the Height Res variable to 4, which (as you'll learn later on) will cause ScanBot to take a lower-quality scan.

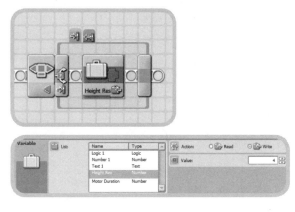

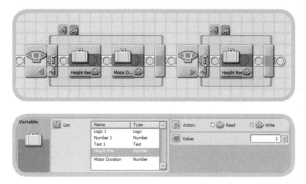

13. Place a Variable block configured to set the Height Res variable to 1 for the high-resolution scan, as shown in Figure 15-13.

Figure 15-10: The Height Res Variable block that determines the scan resolution

11. Add a Variable block configured to set the value of the Motor Duration variable to 40 to help set a low resolution, as shown in Figure 15-11.

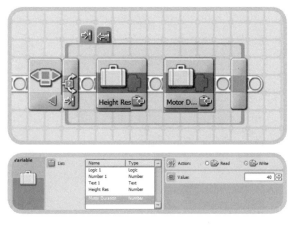

Figure 15-11: The Variable block that also determines the scan resolution

12. Place a Switch block configured to detect when the Right NXT button is pressed, as shown in Figure 15-12.

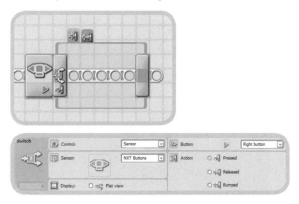

Figure 15-12: The SWITCH block for determining whether the Right NXT button was pressed

Figure 15-13: The Height Res Variable block set for high resolution

14. Place a Variable block configured to set the Motor Duration variable to 16 (for high resolution), as shown in Figure 15-14.

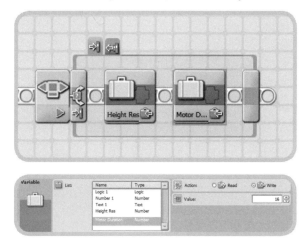

Figure 15-14: The Motor Duration Variable block set for high resolution

15. Now select all the program blocks that you added and turn them into a new My Block called *Resolution Switch*. (Select an appropriate icon too, if you like.) This My Block will determine the resolution of the scan. Now delete the My Block from your workspace so you can make the next one.

the threshold selector my block

The Threshold Selector My Block enables us to use the NXT buttons to specify the *light threshold*—the number that determines whether a given light reading will cause ScanBot to draw a black or a white dot on the LCD. Since the Light Sensor measures only the intensity of light reflected by an area of the picture, not the color of that area, we will use a numerical threshold to compare the measured intensity against to decide whether to draw a black or white dot for that area of the scanned image. If the intensity is higher than the threshold, a white dot will be drawn. If the intensity is lower, a black dot will be drawn. It's important to be able to adjust the value of this threshold to compensate for different lighting effects, shades of gray in the picture, and so on.

1. Place a Variable block configured to read the Light Threshold variable you defined earlier, as shown in Figure 15-15. This block will determine the threshold between black and white, and the user will adjust its value to modify the threshold.

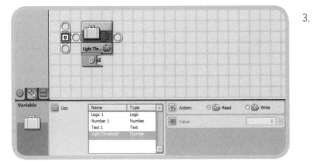

Figure 15-15: The Light Threshold Variable block for determining the threshold between black and white

2. The user needs to see what the light threshold is, so we'll add a Number to Text block that will convert the number from the Light Threshold variable into text so it can be displayed on the LCD (see Figure 15-16).

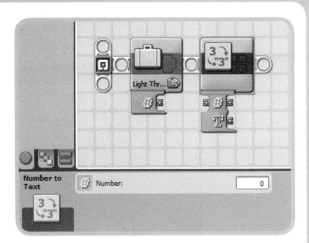

Figure 15-16: The Number to Text block for outputting the numerical value of the Light Threshold variable as text to be displayed on the LCD

3. Place a Text block with the text *Threshold:* (note that there's a space after the colon), so that instead of just displaying the number by itself, this user will see the word *Threshold:* in front of the value (see Figure 15-17).

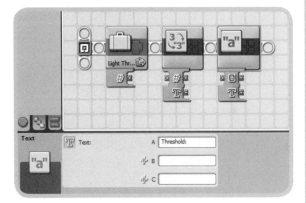

Figure 15-17: The Text block that tells users that they're adjusting the light threshold

4. Place a Display block configured to display text so that both the variable and the Threshold: label will be displayed on the LCD (see Figure 15-18).

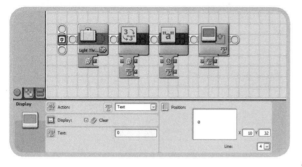

Figure 15-18: The Display block for displaying the threshold variable

5. Wire the four previous blocks (the Variable, Number to Text, Text, and Display blocks), as shown in Figure 15-19.

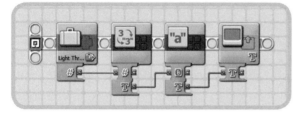

Figure 15-19: The four blocks wired together correctly

6. Place a Loop block configured to detect when the user has finished setting the threshold (see Figure 15-20). It will do this by detecting when the Enter NXT button is bumped.

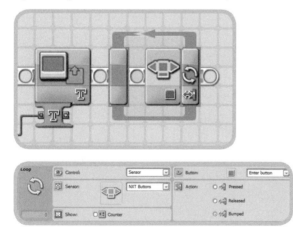

Figure 15-20: The Loop block configured to detect the Enter button being bumped

7. Place a Switch block configured to detect when the Left NXT button is pressed, as shown in Figure 15-21.

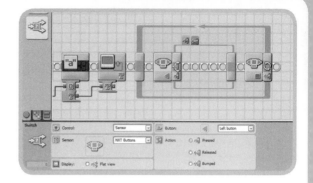

Figure 15-21: The Switch block for detecting the Left button being pressed

8. Place a Variable block configured to read the Light Threshold variable, as shown in Figure 15-22.

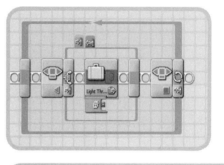

Figure 15-22: The Light Threshold Variable block

9. Place a Math block configured for Subtraction, as shown in Figure 15-23.

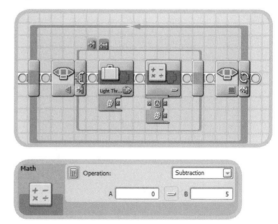

Figure 15-23: The Math block for decreasing the value of the Light Threshold variable

10. Place another Variable block configured to write to the Light Threshold variable, as shown in Figure 15-24.

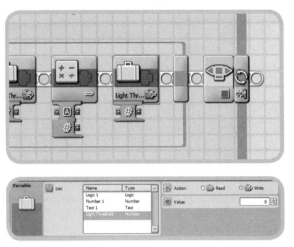

Figure 15-24: The Light Threshold Variable block that will receive the decreased value

11. Wire the three previous blocks together, as shown in Figure 15-25.

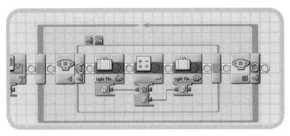

Figure 15-25: The three blocks wired together correctly

With the three blocks wired this way and placed inside the Left button Switch block, the Light Threshold variable will be decreased when the Left NXT button is pressed.

increasing the threshold

Now we have to make the same thing happen when the Right NXT button is pressed, except that we want the Light Threshold variable to be *increased* instead of decreased.

1. Place a Switch block configured to detect the Right NXT button being pressed, as shown in Figure 15-26.

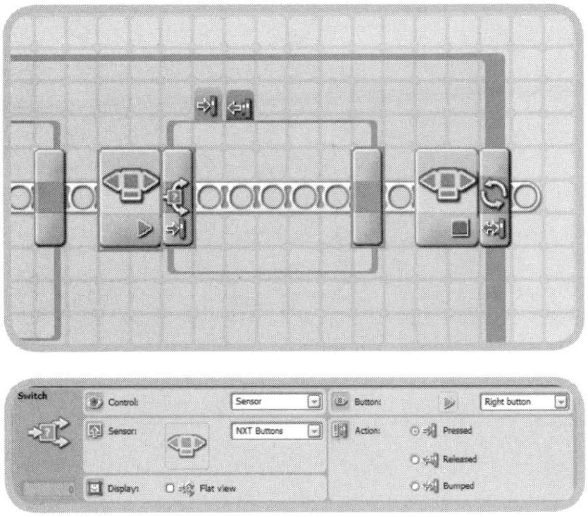

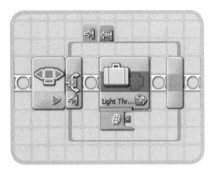

Figure 15-26: The Switch block for detecting the Right button being pressed

2. Place a Variable block configured to read the Light Threshold variable, as shown in Figure 15-27.

Figure 15-27: The Light Threshold Variable block

3. Place a Math block configured to add 5, as shown in Figure 15-28.

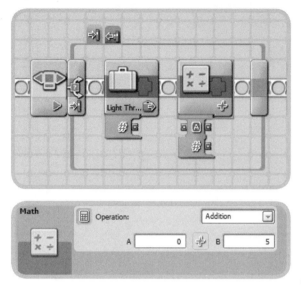

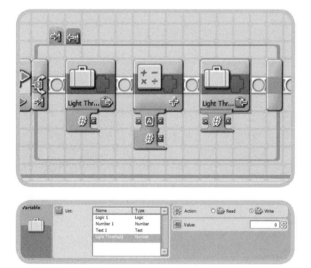

Figure 15-28: The Math block for increasing the value of the Light Threshold variable

4. Place a Variable block configured to write to the Light Threshold variable as shown in Figure 15-29.

Figure 15-29: The Light Threshold Variable block that will receive the increased value

5. Wire the three previous blocks together, as shown in Figure 15-30.

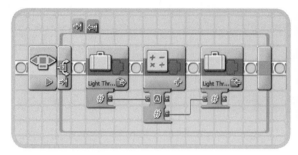

Figure 15-30: The three blocks wired together correctly

With these three blocks wired this way and placed inside the Right button Switch block, the Light Threshold variable will be increased when the Right NXT button is pressed.

displaying the updated threshold

After the light threshold has been adjusted, we need to display the new reading, which we'll do in the same way we displayed it at the start.

1. Place a Variable block configured to read the Light Threshold variable, as shown in Figure 15-31.

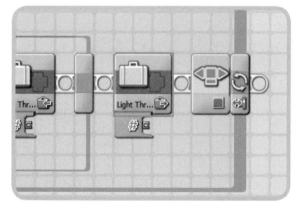

Figure 15-31: The Light Threshold Variable block that will output its adjusted value to be displayed on the LCD

2. Place a Number to Text block, as shown in Figure 15-32.

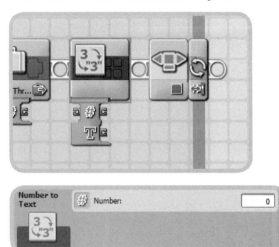

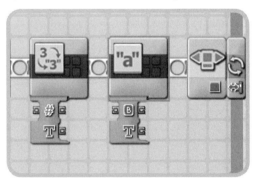

Figure 15-32: The Number to Text block for outputting the numerical value of the adjusted Light Threshold variable as text to be displayed on the LCD

3. Place a Text block with the text *Threshold:* (remember the space after the colon!), as shown in Figure 15-33.

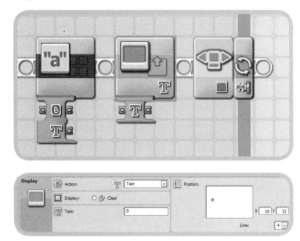

Figure 15-33: The Text block for telling users that they're adjusting the Light Threshold variable

4. Place a Display block configured to display text, as shown in Figure 15-34.

Figure 15-34: The Display block for displaying the adjusted Threshold variable

5. Wire the four previous blocks, as shown in Figure 15-35.

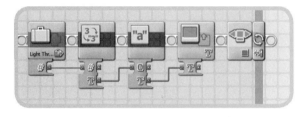

Figure 15-35: The four blocks wired together correctly

Now the LCD will continually update the displayed variable to show any adjustment of it.

6. Place a Wait block configured to wait for 0.2 seconds, as shown in Figure 15-36.

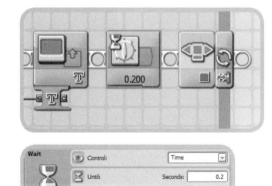

Figure 15-36: The Wait block that will slow down the adjustment rate

This Wait block keeps the light threshold from being adjusted too quickly. Without it, the slightest touch of an arrow key would change the threshold by a huge amount (if you take out the Wait block, you can see this happen).

7. Once again, select all the program blocks you placed and make a new My Block with them. Name it *Threshold Selector*, since it determines the resolution of the scan. Then, as you did before, select an appropriate icon. After finishing, delete the My Block from the workspace so you can make the main program.

main program

This program will make ScanBot live up to its name. First, it enables a user to select the resolution and light threshold of the scan (using the My Blocks that you just made). Then, it turns on a motor, which starts the Light Sensor carriage moving across the bridge. While it's moving, the Light Sensor will take regular readings every few degrees of movement. Depending on the light threshold selected earlier by the user, a black or white dot will be displayed on the LCD each time the Light Sensor takes a reading. Once the Light Sensor carriage has traveled all the way across the bridge, ScanBot will move down a little bit, and the process will repeat until the entire LCD shows a scan of the image.

1. Place a Variable block configured to set the X Position variable you defined earlier to 99, as shown in Figure 15-37. This variable determines the x-coordinate of the dot being drawn on the LCD. Since we want ScanBot to draw the image from right to left, we initially set the X position at the right side of the LCD, which is 99.

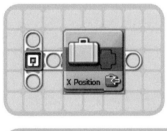

Figure 15-37: The X Position variable set to the correct starting number

2. Place a Variable block configured to set the Y Position variable you defined earlier to 63, as shown in Figure 15-38. This variable determines the y-coordinate of the dot being drawn on the LCD. Since we want ScanBot to draw from top to bottom, we initially set the Y position at the top of the LCD, which is 63.

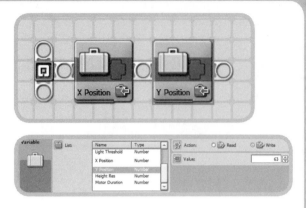

Figure 15-38: The Y Position variable set to the correct starting number

3. Place the Resolution Switch My Block you made earlier as shown in Figure 15-39.

Figure 15-39: The My Block for enabling the user to select the scan resolution

4. Place a Motor block configured to make motor C move for 0.5 seconds, as shown in Figure 15-40. When the scan begins, we want to make sure the Light Sensor carriage is all the way at one end. This Motor block makes sure it's all the way at the right end, as long as you place it close to the end.

Figure 15-40: The Motor block for resetting the Light Sensor carriage to the starting position

5. Place the Threshold Selector My Block that you made earlier, as shown in Figure 15-41.

Figure 15-41: The My Block for enabling the user to select the threshold between black colors and white colors

6. Place a Display block configured to reset the display, as shown in Figure 15-42. This block will reset the display, so nothing will be on the LCD when the scan starts.

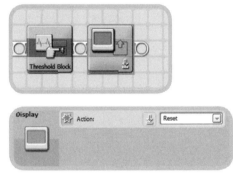

Figure 15-42: The Display block for resetting the display, in preparation for the scan

7. Place a Motor block configured to make motor C move for an unlimited duration, as shown in Figure 15-43.

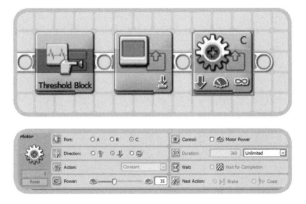

Figure 15-43: The Motor block for making the Light Sensor carriage start moving across the image being scanned

This Motor block makes the Light Sensor carriage move across the bridge at a constant rate.

8. Place a Loop block configured to loop forever, as shown in Figure 15-44. Instead of inserting program blocks to control each pass of the Light Sensor, we can simply program it to make one pass, move down a little, and repeat itself. This makes the program much simpler.

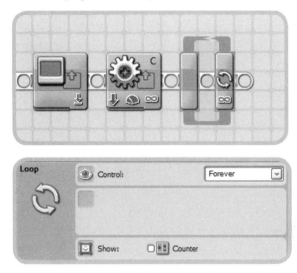

Figure 15-44: The Loop block that will contain the rest of the program blocks

9. Place a Sensor block configured to record the rotations of output C, as shown in Figure 15-45. To be precise, we'll use rotations to determine exactly when to measure the light value of the page, and draw a black or white dot based on that value. Before doing this, however, we need to reset the rotations to zero.

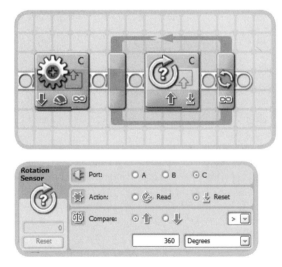

Figure 15-45: The Sensor block for resetting motor C's rotations

10. Place a Wait block configured to wait for six rotations of motor C, as shown in Figure 15-46. This Wait block will make ScanBot record a light reading every 6 degrees.

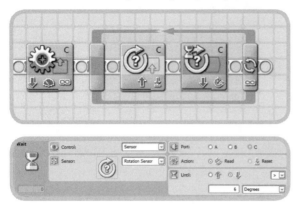

Figure 15-46: The Wait block for determining when to record a light reading

11. Place a Variable block configured to read the Light Threshold variable, as shown in Figure 15-47. This is the variable that, as you will see, tells ScanBot whether to draw a black or white dot on the LCD.

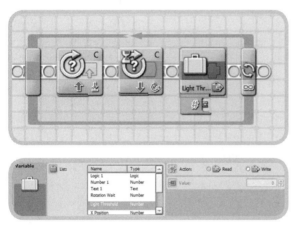

Figure 15-47: The Variable block that determines whether ScanBot draws a black dot or white dot

12. Place a Sensor block configured to record the reading from the Light Sensor, as shown in Figure 15-48. This Sensor block will output the light reading, which we can compare with the Light Threshold variable to determine which color dot to draw.

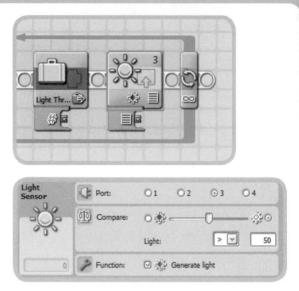

Figure 15-48: The Sensor block for recording the light reading

13. Place a Compare block configured to perform a Less than comparison, as shown in Figure 15-49. This Compare block will be what we use to compare the two values.

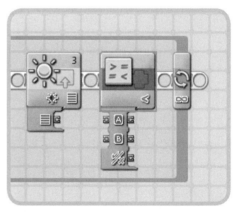

Figure 15-49: The Compare block for comparing the light reading with the Light Threshold variable

14. Place a Switch block configured to detect a Logic signal, as shown in Figure 15-50. This Switch block takes the logic value from the Compare block and uses it to draw either a black or a white dot.

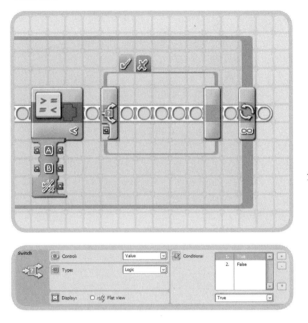

Figure 15-50: The Switch block configured to detect a Logic signal

15. Wire the four previous blocks, as shown in Figure 15-51. With this wiring configuration, the Compare block will take the values of the Light Threshold variable and the Sensor block, compare them, and output a Logic signal (*true* or *false*) based on that comparison. The Switch block will take this signal and use it to either draw a black or white dot on the LCD.

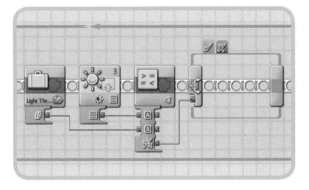

Figure 15-51: The four previous blocks wired together correctly

The NXT-G Display block can draw lines. We'll make use of this feature, since ScanBot draws lines instead of dots. To determine the position of the line, two sets of x- and y-coordinates have to be input: the beginning x- and y-coordinates, and the ending x- and y-coordinates. For the beginning x- and y-coordinates, we simply have the X Position and Y Position variables. For the ending coordinates, we have the same value for x (since the line needs to be straight), but a lesser value than the beginning y-coordinate (so that the line will be longer than one pixel). To do this, we subtract the Height Res variable from the Y Position variable, so the line will be longer or shorter, depending on the value of the Height Res variable. In this way, the user can select the height of the line (and therefore the resolution of the scan) simply by changing the value of the Height Res variable.

16. Place a Variable block configured to read the Height Res variable, as shown in Figure 15-52. This Variable block will determine how tall the line that ScanBot draws will be. (ScanBot doesn't draw one pixel at a time; it draws multiple pixels in a line.)

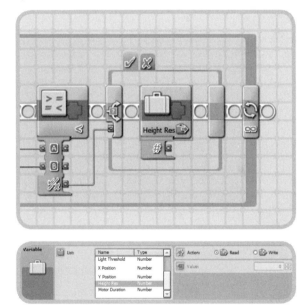

Figure 15-52: The Variable block for determining the scan resolution

17. Place a Variable block configured to read the Y Position variable, as shown in Figure 15-53.

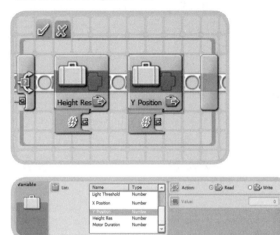

Figure 15-53: The Variable block that also determines the height of the line being drawn

18. Place a Math block configured for Subtraction, as shown in Figure 15-54. This Math block subtracts the Height Res variable from the Y Position variable, thus outputting the number we need for the end y-coordinate.

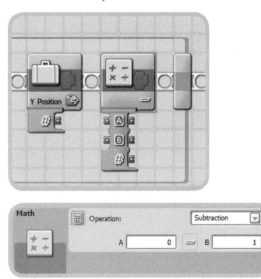

Figure 15-54: The Math block that will subtract Height Res from Y Position

19. Place a Variable block configured to read the X Position variable, as shown in Figure 15-55.

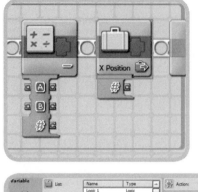

Figure 15-55: The Variable block for determining the end x-coordinate of the line being drawn

NOTE Later, we'll take the numerical output of this variable to use as the end x-coordinate of the line being drawn.

20. Place a Variable block configured to read the Y Position variable, as shown in Figure 15-56. This variable's numerical output will be used for the beginning y-coordinate of the line being drawn.

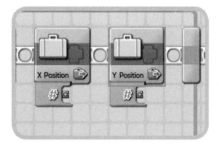

Figure 15-56: The Variable block for determining the y-coordinate of the line being drawn

21. Place a Variable block configured to read the X Position variable, as shown in Figure 15-57. This variable's output will be used as the beginning x-coordinate of the line being drawn.

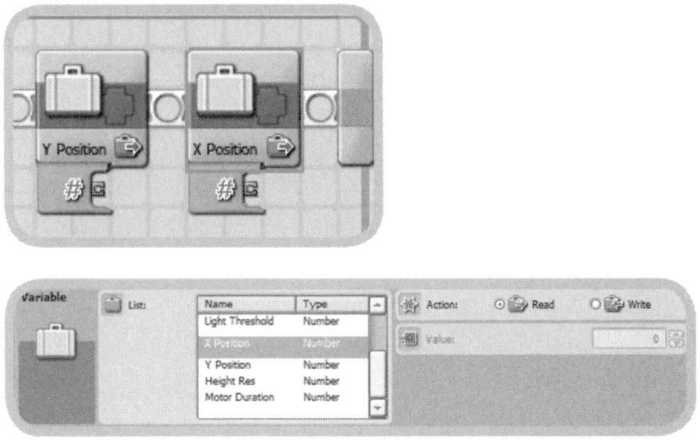

Figure 15-57: The Variable block for determining the x-coordinate

22. Place a Display block configured to display a line, as shown in Figure 15-58. Make sure it's not configured to clear the previous display because it's building onto the picture, and we want to keep the lines that have already been drawn.

Figure 15-58: The Display block for drawing a scanned line on the LCD

23. Wire the seven previous blocks, as shown in Figure 15-59.

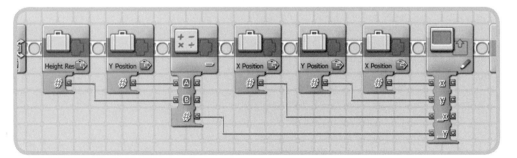

Figure 15-59: The seven previous blocks wired together correctly

24. Place a Sound block configured to play a tone, as shown in Figure 15-60.

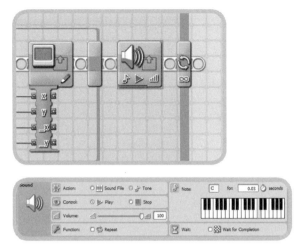

Figure 15-60: The Sound block that indicates ScanBot just took a scan

NOTE Although not at all necessary, you might like having ScanBot play a tone as it makes a scan. Feel free to try different tones, or leave this block out if you don't want it.

25. Place a Variable block configured to read the X Position, as shown in Figure 15-61.

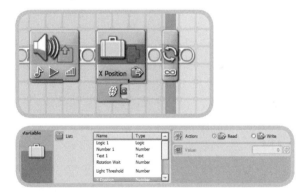

Figure 15-61: The X Position Variable block

26. After drawing a line on the LCD, we need to decrease the value of X Position so that the next line will be drawn to the left of the one just drawn. Place a Math block configured for Subtraction, as shown in Figure 15-62.

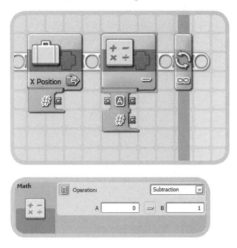

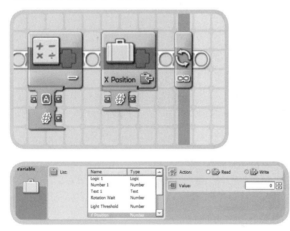

Figure 15-62: The Math block for decreasing the value of X Position

27. Place a Variable block configured to write to the X Position variable, as shown in Figure 15-63.

Figure 15-63: The X Position Variable block for receiving the decreased value

28. Wire the three previous blocks together, as shown in Figure 15-64. These three blocks will now decrease the value of X Position by 1, and therefore make the next line be drawn to the left of the previous one.

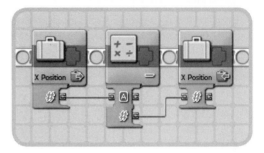

Figure 15-64: The three previous blocks wired correctly

deciding whether to scan the next row

ScanBot scans an image by moving all the way across it, then moving down a little bit, and moving all the way across again. Therefore, after the last line has been drawn at the far left of the LCD, we need to tell ScanBot to move down and move across again. To detect whether the last line was drawn at the far left of the LCD (actually, just close to the far left), we can check to see if the X Position variable is equal to 3 (since that's almost all the way at the left of the LCD). Here's how.

1. Place a Variable block configured to read the X Position variable, as shown in Figure 15-65.

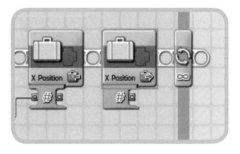

Figure 15-65: The X Position Variable block for determining whether ScanBot should move down the image

2. Place a Compare block configured to perform an Equals comparison, as shown in Figure 15-66.

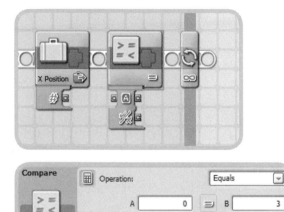

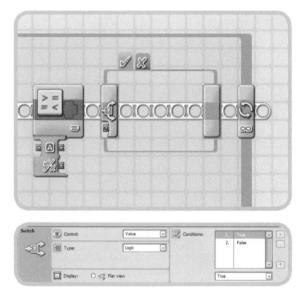

Figure 15-66: The Compare block that determines whether ScanBot should move down and scan across again

3. Place a Switch block configured to detect a Logic signal, as shown in Figure 15-67.

Figure 15-67: The Switch block for making ScanBot move down if X Position equals 3

4. Wire the three previous blocks, as shown in Figure 15-68.

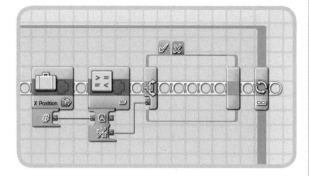

Figure 15-68: The three previous blocks wired correctly

These blocks will now make a series of events (that will be placed inside the Switch block) occur only if X Position equals 3.

5. Since the Light Sensor carriage (powered by motor C) was told to move for an unlimited amount of time, we need to stop it because it's now at the end. To do so, place a Move block configured to stop output C, as shown in Figure 15-69.

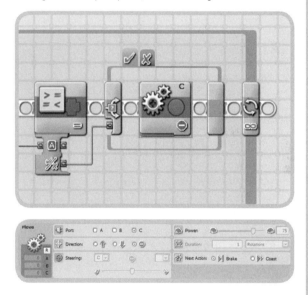

Figure 15-69: The Move block for stopping the Light Sensor carriage

6. Place a Variable block configured to read the Motor Duration variable you defined earlier, as shown in Figure 15-70.

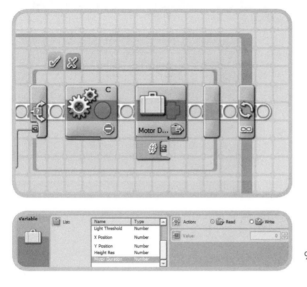

Figure 15-70: The Variable block that determines how far down ScanBot should move

NOTE Remember, this variable was one of the variables whose value depended on the resolution selected. That's because this variable determines how far down ScanBot moves. If the resolution is low, we want ScanBot to move down farther (thus being less precise); if it's high, we want ScanBot to move down a shorter amount (thereby being more precise).

7. Place a Motor block configured to make output B turn 16 degrees, as shown in Figure 15-71.

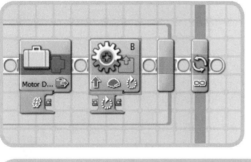

Figure 15-71: The Motor block for moving ScanBot down

8. Wire the two previous blocks, as shown in Figure 15-72. These two blocks will make ScanBot move down for a certain amount, depending on the resolution selected at the beginning of the program.

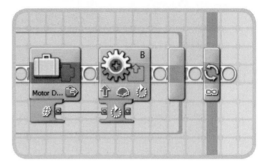

Figure 15-72: The two previous blocks wired correctly

9. Place a Motor block configured to make output C move for 1.7 seconds, as shown in Figure 15-73. This block will make the Light Sensor carriage go back to its starting position so it can make another pass across the image.

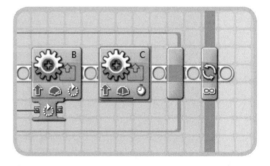

Figure 15-73: The Motor block for resetting the Light Sensor carriage to its starting position

After ScanBot moves down, we need to decrease the y-coordinate so the line being drawn will correspond with the part of the image being scanned.

Place a Variable block configured to read the Height Res variable, as shown in Figure 15-74.

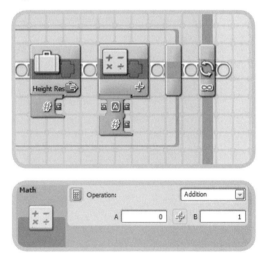

Figure 15-74: The Variable block for decreasing the y-coordinate

10. Place a Math block configured for Addition, as shown in Figure 15-75.

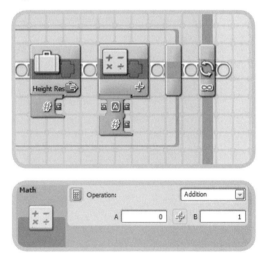

Figure 15-75: The Math block for increasing Height Res by 1

NOTE As you might have noticed, we can't just subtract Height Res by itself, or the next line would overlap the previous line by one pixel. First we need to add 1 to Height Res, so that the new line will be drawn directly under the previous line without any overlap.

11. Place a Variable block configured to read the Y Position variable, as shown in Figure 15-76.

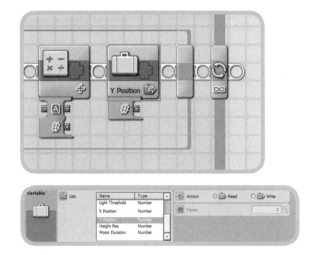

Figure 15-76: The Variable block that will be decreased

12. Place a Math block configured for Subtraction, as shown in Figure 15-77.

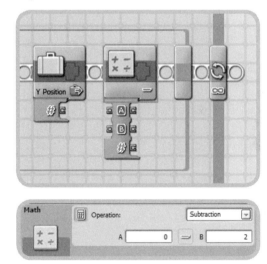

Figure 15-77: The Math block that will decrease the Y Position variable

13. Place a Variable block configured to write to the Y Position variable, as shown in Figure 15-78.

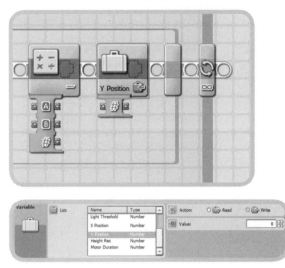

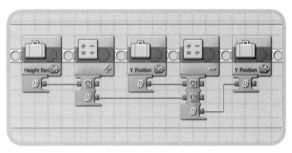

Figure 15-78: The Variable block that will receive the decreased value

14. Wire the five previous blocks, as shown in Figure 15-79.

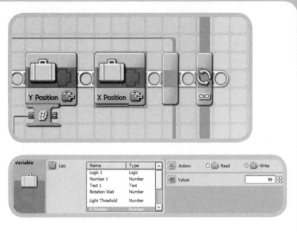

Figure 15-79: The five previous blocks wired correctly

These blocks will make the next line be drawn directly below the previous lines.

Since ScanBot is making another pass across the image from the right side, we need to reset the X Position variable to 99, so the lines will be drawn from right to left again. To do so, we place a Variable block configured to set the X Position variable to 99, as shown in Figure 15-80.

Figure 15-80: The X Position Variable block's numerical value being set back to 99

15. Place a Wait block configured to wait for 1 second, as shown in Figure 15-81. This Wait block ensures that any bouncing or vibrating that might occur when ScanBot moves down has stopped, so that the scan will be more precise.

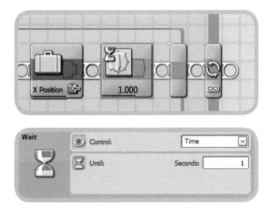

Figure 15-81: The Wait block configured to wait for 1 second

16. Place a Motor block configured to make motor C move for an unlimited duration, as shown in Figure 15-82. This block makes the Light Sensor carriage start moving across again to scan another row of the image.

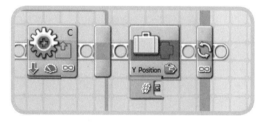

Figure 15-82: The Motor block for making the Light Sensor carriage start moving across again

17. After ScanBot draws lines on the bottom of the LCD, it's done scanning and should let the user see the image. If the Y Position is less than zero, we know this is the case. Therefore, we just need to see whether the Y Position is less than zero. To do this, we place a Variable block configured to read the Y Position variable, as shown in Figure 15-83.

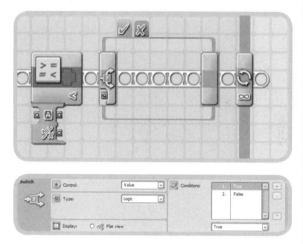

Figure 15-83: The Variable block that determines whether ScanBot is done scanning the image

18. Place a Compare block configured to perform a Less than comparison, as shown in Figure 15-84.

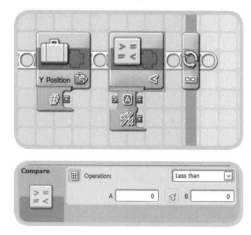

Figure 15-84: The Compare block that also determines whether the scan is complete

19. Place a Switch block configured to detect a Logic signal, as shown in Figure 15-85.

Figure 15-85: The Switch block that stops the program if the scan is complete

20. Wire the three previous blocks, as shown in Figure 15-86. These blocks will make a series of events (which will be placed inside the Switch block) occur only if the Y Position is less than zero.

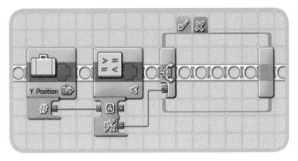

Figure 15-86: The three previous blocks wired correctly

21. Place a Move block configured to stop outputs C and B, as shown in Figure 15-87. Since the scan would be complete if the blocks in this switch are being activated, this Move block makes ScanBot stop moving and enables the user to view the scanned image.

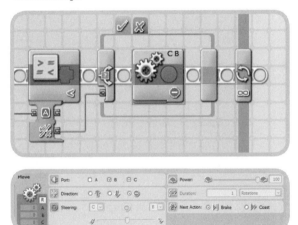

Figure 15-87: The Move block that will make ScanBot stop moving

22. If we were to simply end the program at this point, the scanned image would disappear (which probably wouldn't be very interesting). Instead, we add a Wait block that delays the display of the image until the user presses the Enter button. Place a Wait block configured to wait for the Enter button to be pressed, as shown in Figure 15-88.

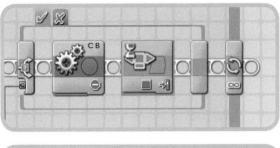

Figure 15-88: The Wait block to enable the user to view the image

23. Place a Stop block, as shown in Figure 15-89. After the user presses the Enter button, this block will end the program.

Figure 15-89: The Stop block that ends the program

Great job—you've finished programming ScanBot, and now you can start scanning!

using ScanBot

After downloading the program to ScanBot, do the following:

1. Place your robot over a piece of paper with an image you want to scan. Make sure the robot is on a smooth, hard surface. (A kitchen countertop or a wooden table works well.)

2. Set the Light Sensor carriage near the right side of the bridge (the side with the motor) and run the program. You should see the options for Preview and High-Res appear, just as you programmed them to do.

3. Select the option you want and press the Enter button. Now you should see *Threshold: 0* appear on the LCD.

4. Press the Right button to increase or the Left button to decrease the threshold. (Try selecting 45 for the threshold—that usually works best. After the scan, you can see whether the threshold needs to be adjusted.)

NOTE If there's too much black on the LCD, the threshold is probably set too high. If, on the other hand, too little of the image is being picked up by ScanBot, the threshold is probably too low.

After you've selected your desired light threshold, press the Enter button, and the Light Sensor carriage should start moving across the bridge. (It will play a tone if you chose to keep that Sound block earlier.) If you look at the LCD as the carriage moves, you should see a line being drawn across the LCD if the Light Sensor is moving over a black area of the picture. That line is the beginning of your scanned image.

After ScanBot finishes, it will stop, and you can view the scan. You can even detach the cables and remove the NXT brick from the main robot to show the scanned image to your friends. When you're done viewing the image, press the Enter button, and the program will end (erasing the image).

trouble-shooting tips

Here are some possible solutions to a few problems you might encounter. Remember, ScanBot works best on smooth, hard surfaces.

* If you're getting black splotches on the LCD and changing the threshold doesn't fix the problem, make sure that the paper is perfectly smooth without any bumps. A good way to do this is to tape the four corners of the paper onto the surface it's on.
* If the Light Sensor carriage gets stuck on the bridge as it goes across, try pressing the top and bottom beams in the bridge together tightly; they shouldn't be uneven. If this doesn't fix it, make sure the cables aren't hindering the movement of the Light Sensor carriage. If neither of these is the problem, try replacing the batteries.
* If your robot isn't working as it should, go back over your program to make sure that it matches the program instructions here. Even the slightest difference in the program could make ScanBot act much differently.

above and beyond

Now that you've built a working scanner, how about experimenting a bit? For example, can you make a robot that can scan an image and copy it to another piece of paper with a marker or pen? Or can you find a way to let a user zoom in or out of a scanned image on the LCD? Whatever your ideas are, go ahead and try making them—after all, that's what LEGO MINDSTORMS NXT robots are all about!

marty: a performance art robot

The audience is captivated by something, but you're still not close enough to see what all the fuss is about. There's no way that you could've missed the crowd gathering excitedly at the edge of the market, and you're keen to see what's going on in there. As you push into the throng and stretch up to see over the people in front of you, you glimpse something racing past, then a line, some curves . . .

With an ordinary marker pen, Marty is a robot capable of creating original pieces of geometric art before your very eyes. This chapter includes building instructions for Marty as well as example programs to enable Marty to create a range of shapes, including polygons, stars, zigzags, and spirals.

The basic concept of Marty is a robot that can move across a surface, drawing as it goes (see Figure 16-1). Whereas a conventional printer or plotter is limited by the size of paper that it can hold, Marty can potentially work with any sized drawing surface.

As Marty travels around the surface, each movement becomes a line or a curve. A series of movements can become anything from a mechanized scribble to a robotic work of art, depending on the sophistication of Marty's program and your imagination!

In addition to the parts from the LEGO MINDSTORMS NXT set required for building Marty, you will also need a drawing surface (e.g., paper on a hard flat surface or a whiteboard), marker pens, and a couple of rubber bands.

design challenges

Marty was designed as a simple two-wheeled chassis with a central module to hold a marker pen. As you work through these building instructions, notice that the section that holds and moves the pen within Marty, the *pen housing*, is a separate module that sits within the rest of the chassis.

holding the pen

The greatest concern I had when designing Marty was ensuring that the marker pen would be held in the correct position for making tight turns. Unless the pen is held correctly, the corner becomes a curve.

To see what I mean, try attaching a pen to the front of a typical two- or three-wheeled robot (e.g., TriBot, the first robot described in the building instructions that come with the NXT kit) and then program it to move forward, turn on the spot, and then go forward again. The solution to this problem is to position the pen exactly halfway between the wheels.

Figure 16-1: Marty, a performance art robot

Related to this was the challenge of finding an effective method for attaching the marker pen to the robot. I knew from previous experience with RIS-based drawing robots that it was important to be able to manually adjust the height of the pen. If the pen is down too far, the wheels will be lifted slightly off the surface, which places the pen too high to be able to draw a solid line. I was also very keen for the robot to be able to lower and raise the marker pen automatically.

Ultimately, I designed the pen grip as a removable module to make its development a little easier and to keep open the possibility of replacing it with an alternative module at a later date—for example, a pen grip that would support marker pens of different colors.

wheels

In addition to the two wheels required for movement, I considered whether to use a caster or two for balance. I decided against using casters, however, because of the need to keep turns tight and straight lines as straight as possible. (Because casters trail the movement of the robot, they can often end up "sideways" after a turn. When the robot moves forward from this position, the robot will tend to wiggle a little as the casters straighten up.) Instead of using a caster, I decided that skid plates or an equivalent would work fine on drawing surfaces such as paper or a whiteboard.

After many successes and failures, I came up with the robot you see in Figure 16-1.

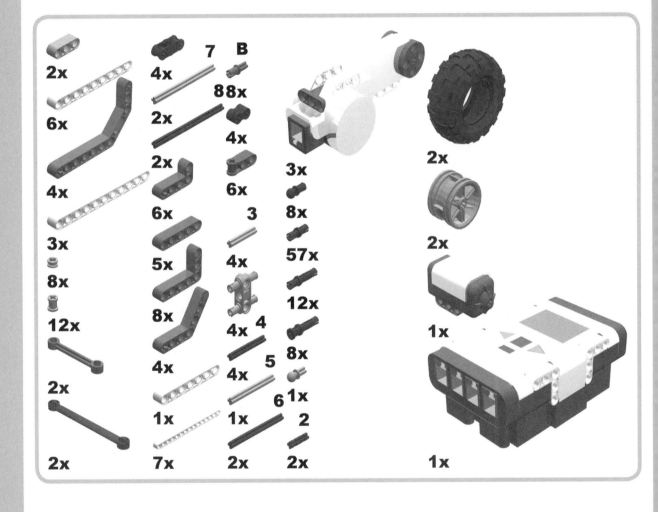

building marty

Building Marty involves the construction of two main parts: a two-wheeled chassis and a pen housing. Both the chassis and the pen housing are presented here as a series of subassemblies to hopefully make building Marty fairly manageable. To build Marty, you will build each of the two main parts separately and then finish off by putting them together.

left motor subassembly

The purpose of the left motor subassembly (shown below) and its mirror image, the right motor subassembly, is to provide structure that includes the motors and wheels and the support "legs" for balance.

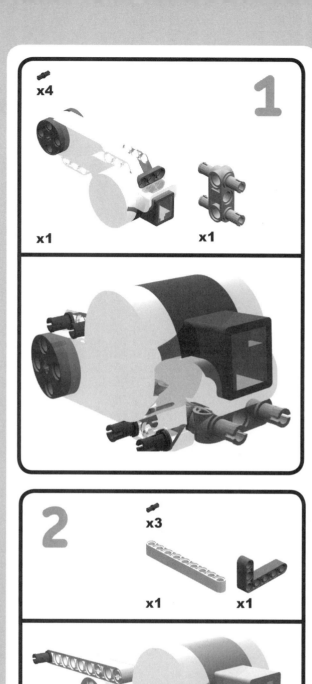

3

x2

x1

x2

4

x2

2

x2

x1

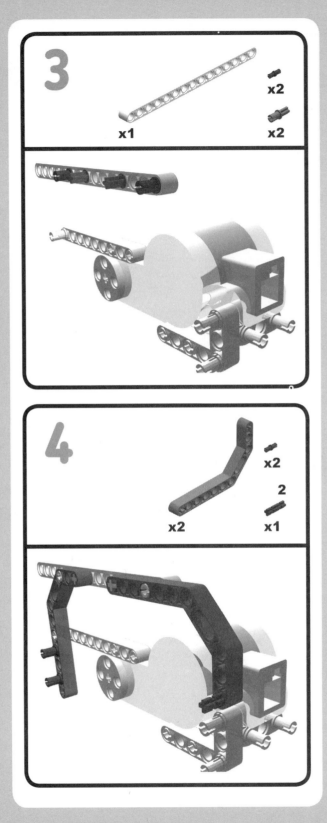

NOTE If you are having trouble fitting the bent beams onto the blue axle pins, it might help to fit the axle side of the pins first.

5

x2

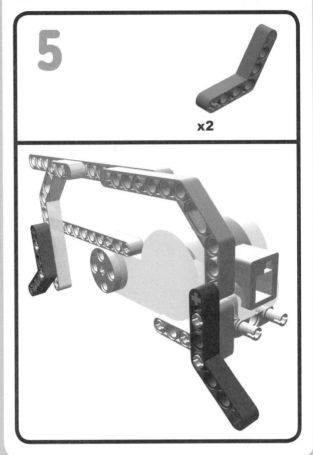

6 x2
x2

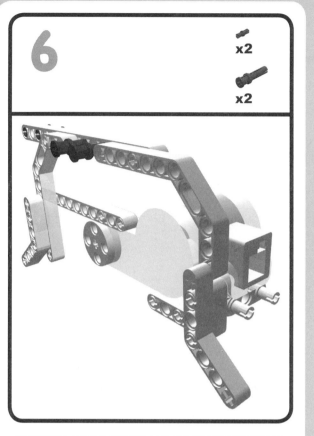

7 x1 x1 6

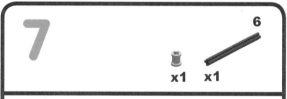

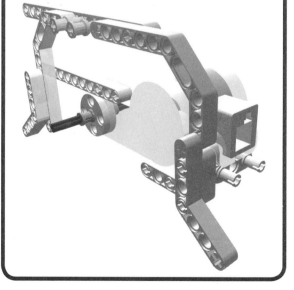

8 x1 x1

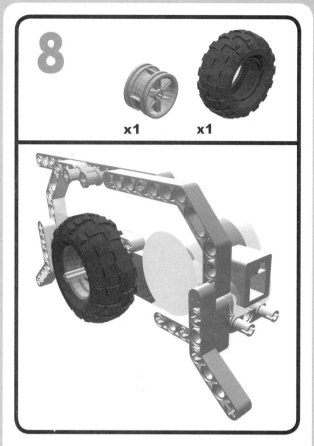

Make sure that the side of the wheel with the star pattern is toward the motor. (This probably won't be a problem because you will find that the wheel doesn't fit properly the other way around, anyway.)

right motor subassembly

As indicated previously, the right motor subassembly (shown below) is the mirror image of the left motor subassembly.

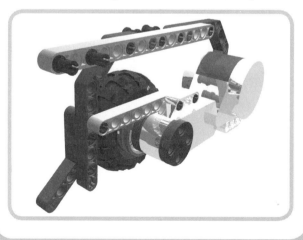

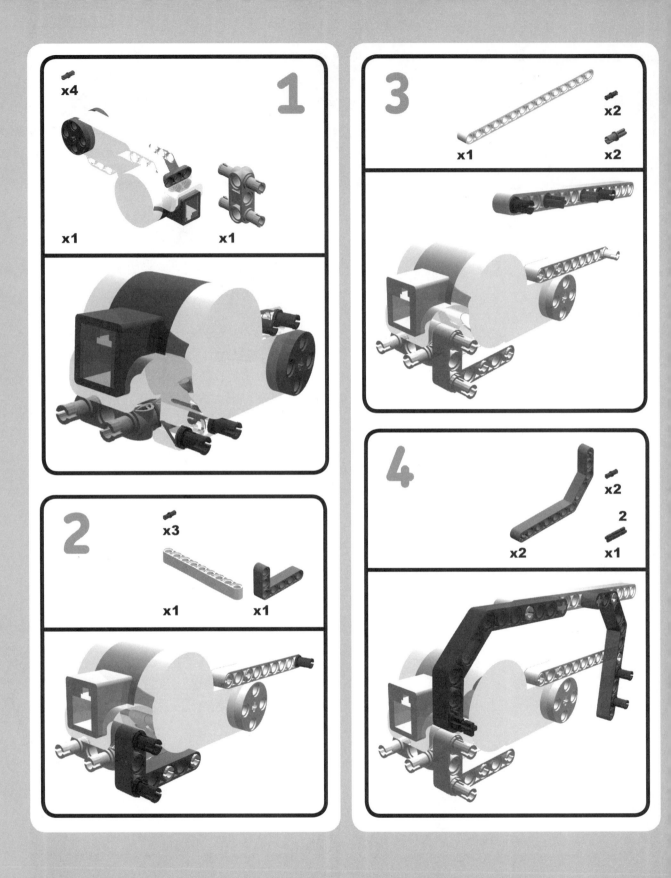

5

x2

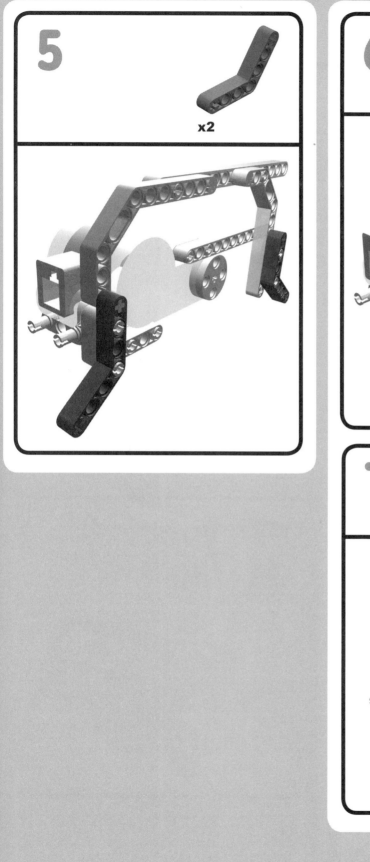

6

x2

x2

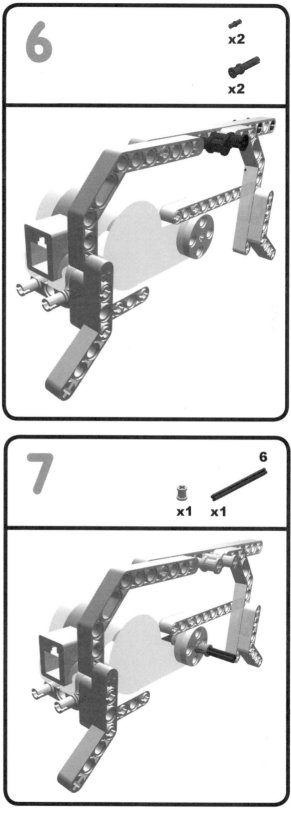

7

6

x1   x1

8

x1

x1

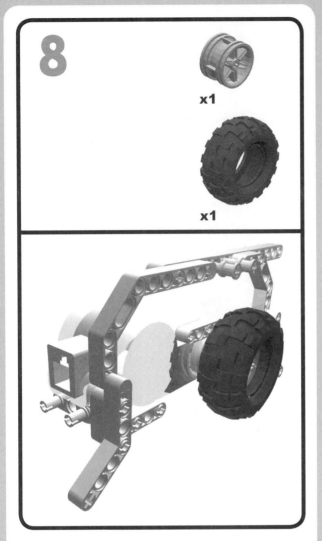

Remember to have the side of the wheel with the star pattern toward the motor.

the two-wheeled chassis

Before moving on to the pen housing, you can combine the two motor subassemblies to make the basic two-wheeled chassis by fitting them onto each side of the NXT brick at the front and connecting by a beam at the rear. An extra pair of beams will also be attached to the NXT brick in preparation for the pen housing that follows.

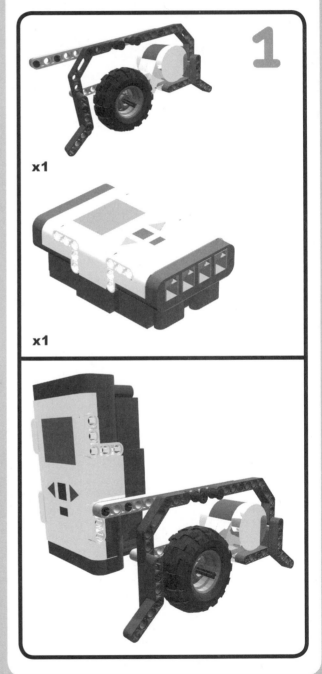

1

x1

x1

2

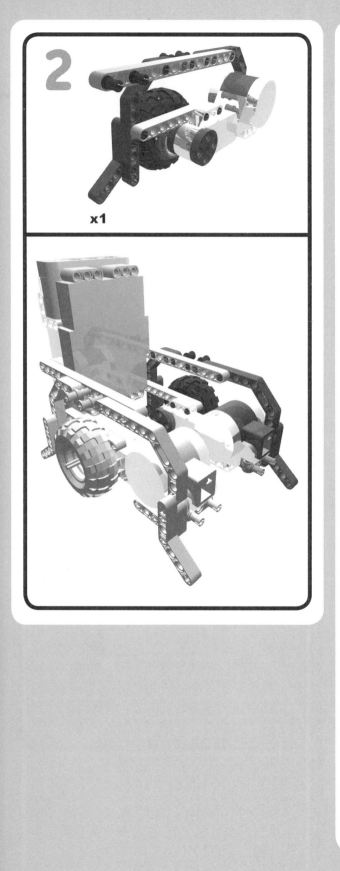

x1

3

x4 x1

4

x4 x2

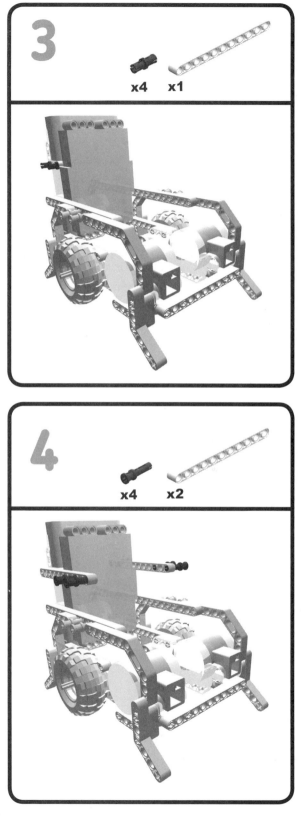

pen grip

As its name suggests, the pen grip (shown here) is the part of Marty that actually holds the pen, and as such it forms a significant part of the pen housing. When Marty is in operation, it is the pen grip that will be moved vertically to raise and lower the pen.

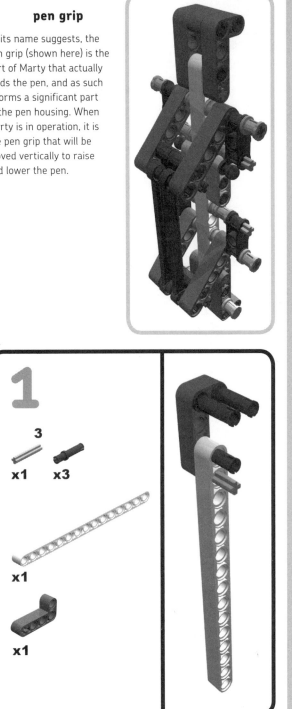

2

x3

3
x1

x1

1

3
x1   x3

x1

x1

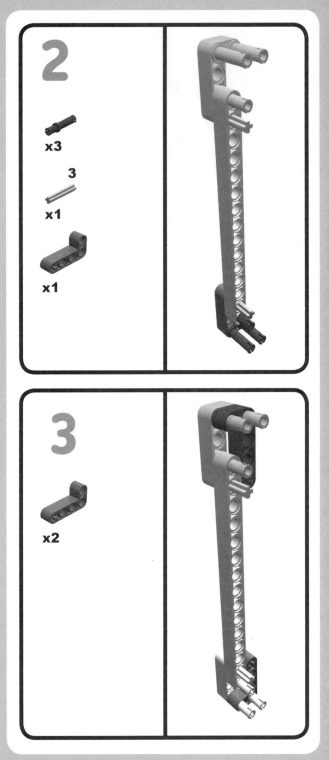

3

x2

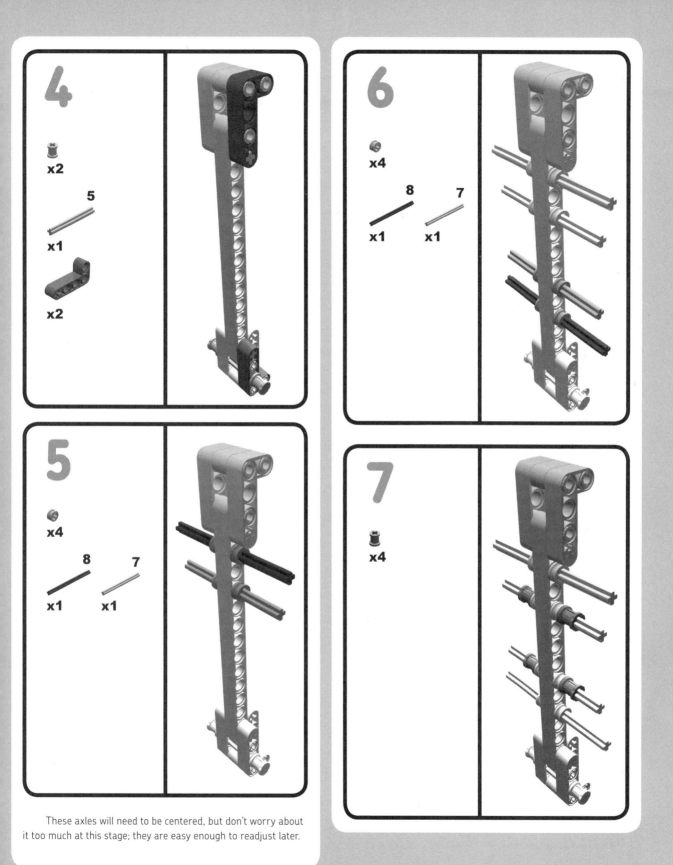

4

x2

5

x1

x2

5

x4

8

x1

7

x1

These axles will need to be centered, but don't worry about it too much at this stage; they are easy enough to readjust later.

6

x4

8

x1

7

x1

7

x4

8

x2

x1

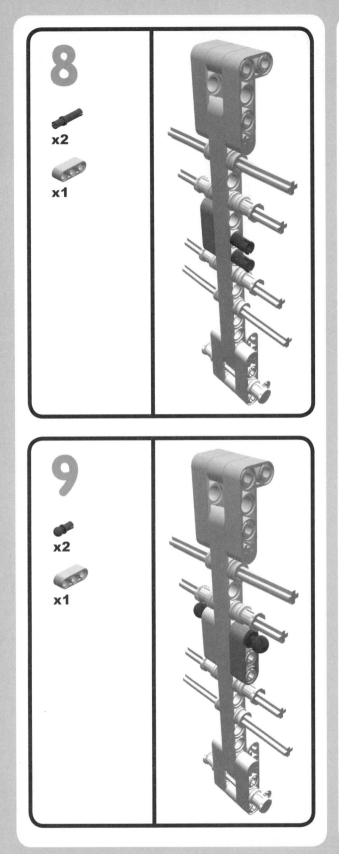

9

x2

x1

10

x2

x2

x1

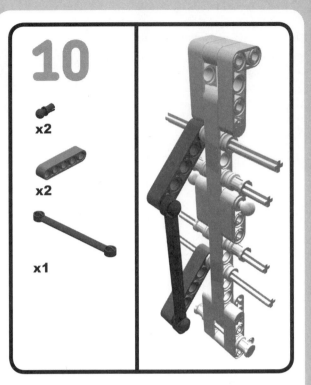

If the link arm added in this step is parallel to the length-15 beam, then everything is going well. If not, then you will need to check the position of all the axles and connecting pegs already attached to the length-15 beam before proceeding. It would also be worth double-checking that the link arm and length-5 arms have been positioned on the correct side of the pen grip assembly.

You will need two of these.

11

x2

4

x2

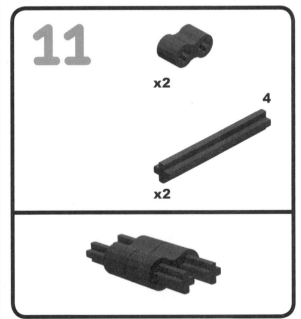

12 x1

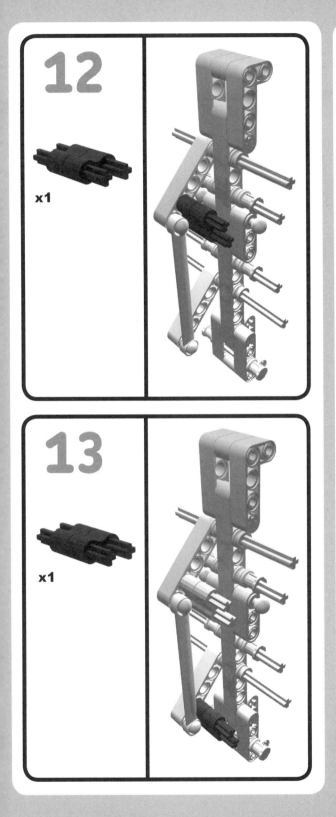

You will need to add a rubber band around each of the friction pieces. This is a bit fiddly, but should end up looking like the two photos here.

Upper friction piece with rubber band

13 x1

Lower friction piece with rubber band

14

x2

x2

x1

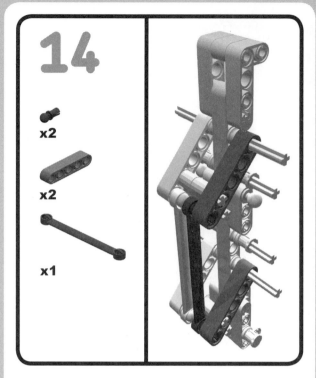

This is another fiddly bit, but the good news is that it becomes easier from here.

15

x4

x4

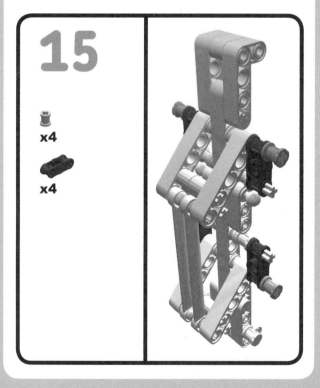

left rail

Put the pen grip to the side for the moment and build the left rail (shown here) and right rail. The purpose of these rails is to hold the pen grip in place within the pen housing and to restrict the pen grip to a vertical movement.

1

x2

x1

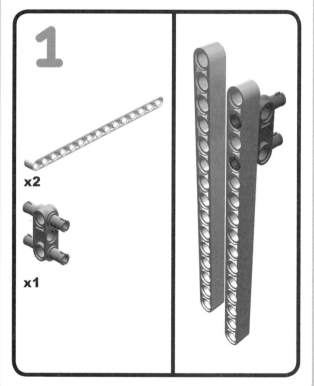

2

x6

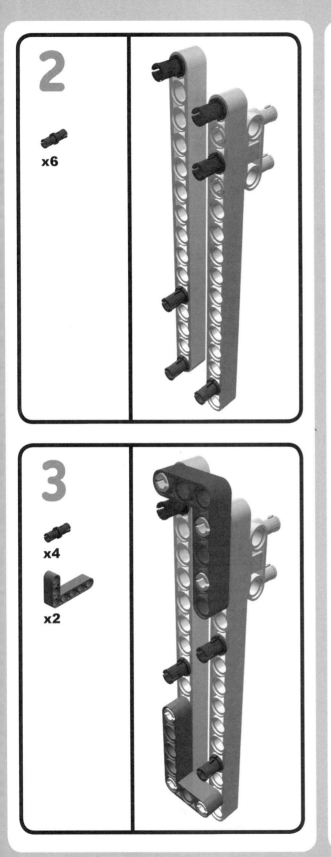

3

x4

x2

4

x2

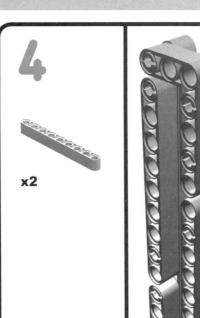

right rail

This is a mirror image of the left rail.

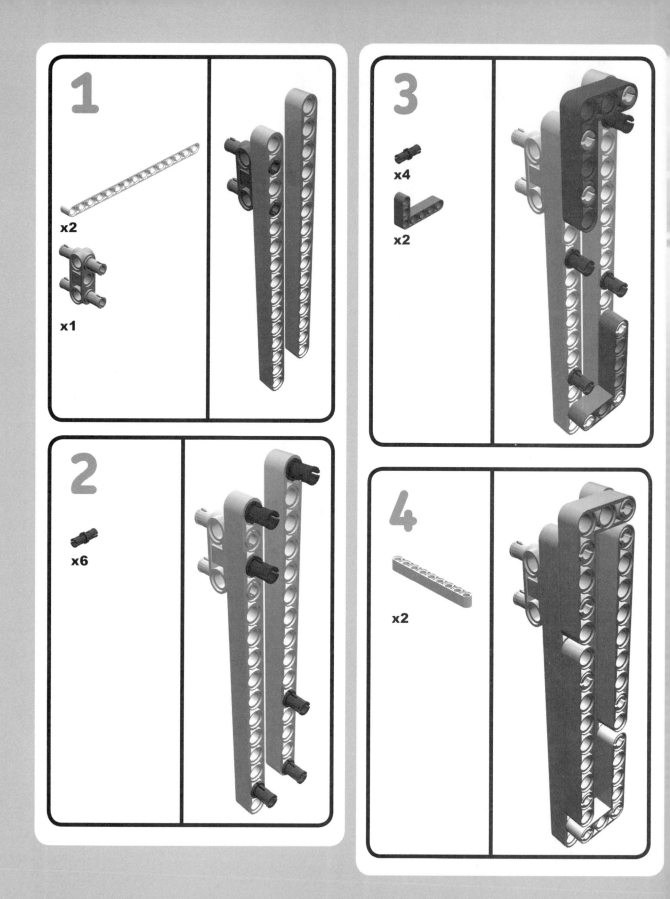

pen housing

Marty finally comes together with the assembly of the pen housing (shown here). Once completed, this module will be dropped into the center of Marty, allowing him to hold a pen and draw.

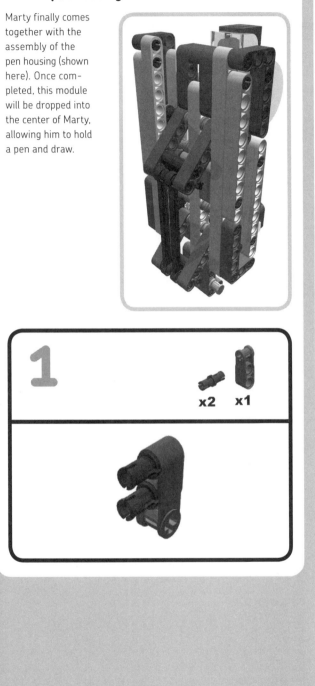

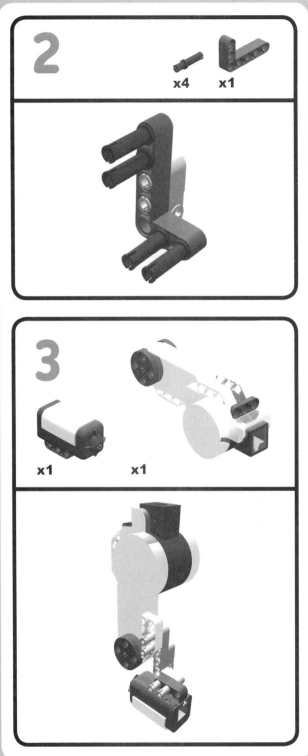

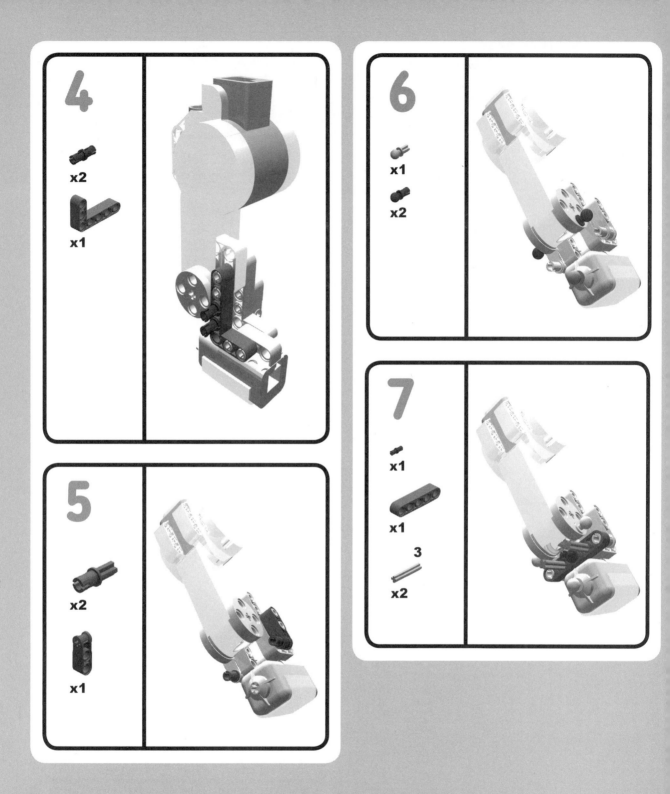

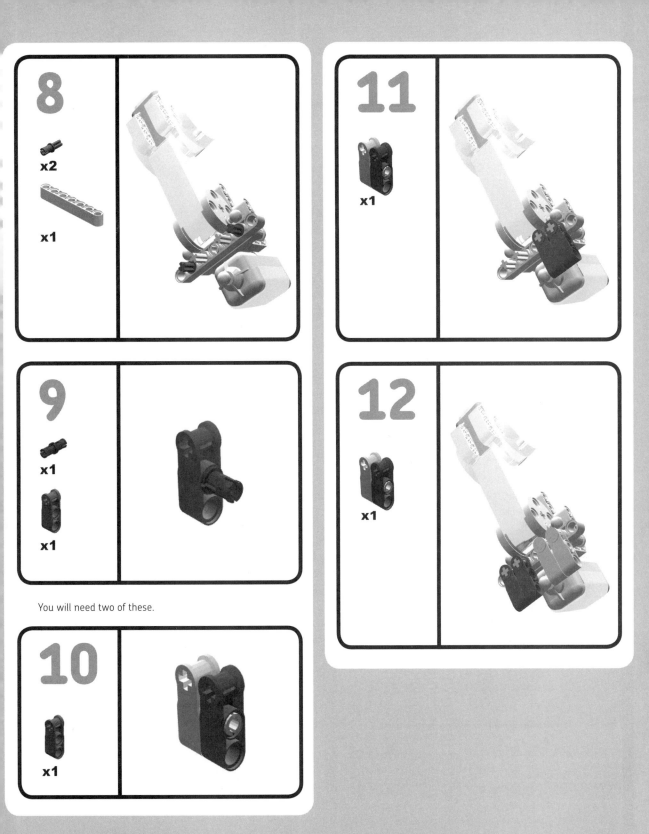

8 x2 x1

9 x1 x1

You will need two of these.

10 x1

11 x1

12 x1

Okay, so this is another fiddly bit, but hopefully you are starting to see what is going on now. You will need to hold the pen grip against the "front" of the pen housing and maneuver it until you can fit the steering links on each side.

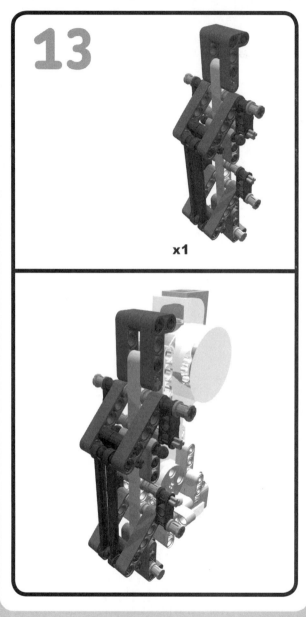

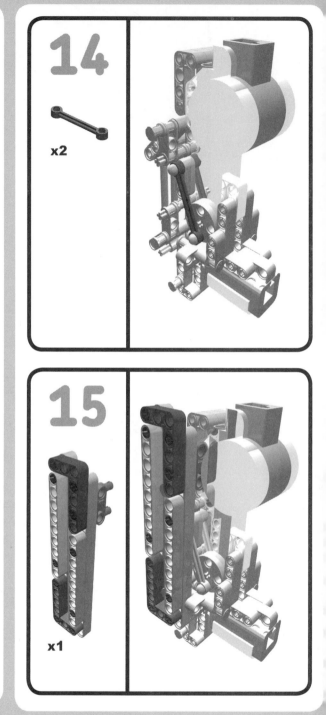

While holding the pen grip against the pen housing, you should be able to fit each of the side rails onto the housing. You will notice that the rails are attached only to the housing at the top. This provides some give at the bottom that allows the Touch Sensor to work correctly.

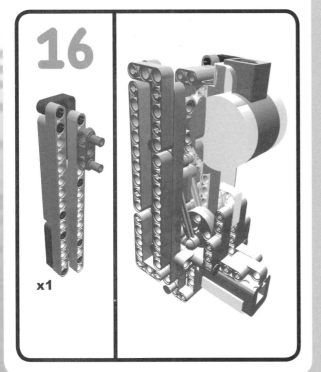

putting it all together

You will notice when inserting the pen housing that the side rails rest nicely on the motors. Line up the pen housing rails with the stop bushing pins, and then press in the pins on both sides to secure the pen housing.

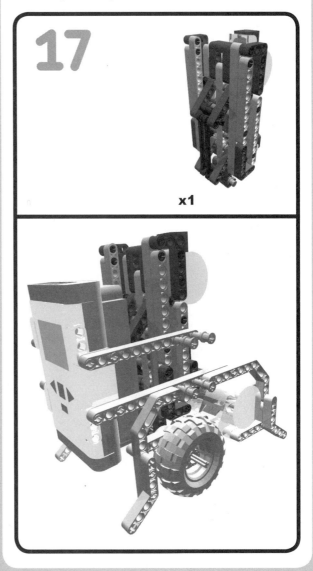

You will need four medium-length wires to attach the motors and Touch Sensor to the NXT brick, as indicated in Table 16-1.

table 16-1: attaching the sensor and motors

NXT port	motor/sensor
Input 1	Touch Sensor
Output A	Pen housing motor
Output B	Right motor
Output C	Left motor

After all the parts have been assembled, you can give Marty a marker pen (as shown in Figure 16-2) so that he can start drawing!

NOTE You might find that it is easier to insert a pen from underneath. Remember that if the tip of the pen is too far below the wheels, the wheels will be lifted off the surface, and the robot will not turn properly. You might need to experiment with the height of the pen, but you will probably find that the tip of the pen needs only very slight pressure onto the surface to work effectively.

Figure 16-2: A view from underneath

You will also soon discover that if the pen is down while you are working on your program or otherwise distracted, the ink will tend to bleed. To avoid bleeding, remember to put the cap back on the pen, or at least tip the robot sideways. Before long, though, you'll find that it's just as easy (and much cooler) to program Marty to automatically raise the pen at the end of each drawing.

programming marty

Although this section will describe how to program Marty to produce a range of shapes—and even some original creations—my main aim is to show you how to write your own programs to create some original drawings. To this end, we will put together a collection of My Blocks to take care of some of Marty's most common functions and then look at a few of the ways in which they can be used. (For an introduction to My Blocks, see "My Blocks Save Time and Simplify Your Programs" on page 16.) First, though, let's start with something a bit simpler.

marty's first drawing

For Marty's first drawing, we will use a Loop and some Move blocks to create a program to sketch a triangle (see Figure 16-3).

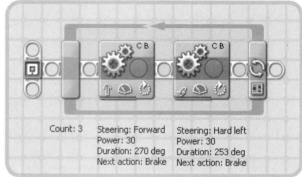

Count: 3   Steering: Forward   Steering: Hard left
Power: 30   Power: 30
Duration: 270 deg   Duration: 253 deg
Next action: Brake   Next action: Brake

Figure 16-3: Triangle

NOTE In this chapter, I assume that you are reasonably familiar with the blocks in the Common Palette, such as Move, Loop, and Wait. For more help with the basics of programming, please see Chapter 2.

NOTE Instead of showing you the settings for every block in these programs, comments below the blocks list significant settings, as shown in Figure 16-3. Although you might need to adjust the settings within each block to match, you won't need to copy the comments as well.

NOTE Remember that the images on the blocks change to reflect their settings. For example, Figure 16-3 shows that the direction of the first Move block is set to forward, whereas the second Move block is set to a sharp left turn.

After you finish your program, set up Marty with a pen pressing slightly onto a piece of paper; then start the program and see what happens. If Marty draws something like the example shown in Figure 16-4, you are off to a great start. If not, there are a number of adjustments that can be made to both the program and the robot. We'll address many of them as we progress through this section. (See the "Troubleshooting" section on page 329 for a summary.)

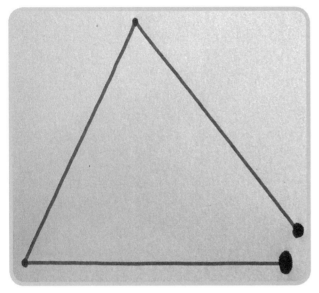

Figure 16-4: Marty's first drawing

NOTE Remember that until you've programmed the pen to lower and raise, you will need to put the cap back on, or at least tip the robot sideways, to avoid the ink bleeding through the paper.

tweaking the program

Although this is a relatively simple program with some significant limitations, it is amazing what you can do with it. By simply playing with the settings for each of the blocks you can have Marty draw a range of shapes.

If you haven't already done so, try changing the settings in this program to see what happens. For example, what happens if you change the number of degrees in the first Move block? What about the second? Can you program Marty to draw a square? What about a star? And so on.

After playing with this program for awhile, you will start to notice some of its limitations, not the least of which is that it is fairly tricky to produce a specific shape such as a triangle or a square on the first attempt. If you have tried to make a shape such as a square or a pentagon, you know that to make the last side of

the shape join up with the first, you need to experiment with the duration of the turn to produce the necessary angles.

Through trial and error I have found that setting the second Move block to 253 degrees results in Marty turning through approximately 120 degrees (to produce a drawn angle of 60 degrees). (Of course, it would be much easier to be able to type in 120 degrees instead of having to find 253 by trial and error.) Likewise, for a square we need to find a duration that makes the robot turn through 90 degrees, and so on, for any regular polygon.

NOTE It would, of course, be much better to have Marty turn exactly 120 or 90 or whatever number of degrees you want. This is addressed in "Marty's Basic My Block Toolkit" below.

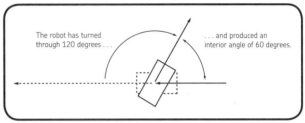

The robot has turned through 120 degrees . . .   . . . and produced an interior angle of 60 degrees.

Similarly, we can't instruct our robot to go forward a particular distance directly. Instead of specifying degrees or rotations, it would be desirable to have Marty move forward (10 cm, for example).

marty's basic my block toolkit

By now, it should be clear that programming Marty will be much easier with a set of My Blocks to take care of some common movements, such as lowering and raising the pen, moving in a straight line a specified distance, and turning on the spot a set number of degrees.

To make creating and using each of the following My Blocks a little easier, I gave each one a name, starting with *mb-*, and an example icon. You might find it easier to follow the example programs later in this chapter if you make similar-looking icons for your My Blocks.

NOTE Remember that when copying the following examples, the comments below each block are only there to make it easier for you to give your blocks the same settings. The comments themselves do not need to be reproduced.

mbPenDown

The purpose of the mbPenDown My Block (see Figure 16-5) is to move the pen grip down until the Touch Sensor is released. When the marker pen is positioned correctly within the pen grip, the Touch Sensor will be released as the pen makes good contact with the drawing surface. If the pen grip is already down, this My Block will appear to do nothing.

NOTE If the pen is set too low, then the pen grip mechanism might actually lift Marty slightly off the drawing surface, preventing the Touch Sensor from being released. To fix this, slightly move the pen up within the pen grip.

The small time delay that occurs after the motor is stopped is designed to ensure that the pen has come to rest before the next command is given. This helps to make sure that line being drawn doesn't appear to "fade in."

mbPenUp

The mbPenUp My Block (see Figure 16-6) is virtually identical to mbPenDown, except that here we're waiting for the Touch Sensor to be pressed instead of released, indicating that the pen grip is up, and the pen is (presumably) off the drawing surface. We've moved the delay to the beginning of the sequence to ensure that the robot has stopped moving before the pen is lifted.

mbForward

It would be nice to be able to move Marty forward by a certain number of centimeters instead of by degrees, rotations, or some measure of time. To do this, we need to determine how many degrees the wheels need to turn for the robot to move forward (1 cm, for example).

One way to do this is to use a formula to calculate the circumference of the tires based on their diameter. As it happens, the diameter of the tire that comes with the NXT kit is 56 mm. (It says so on the side of the tire.) We can use this formula to determine the circumference of the tire:

$$\text{circumference} = \pi \times \text{diameter}$$

Since the diameter is 56 mm, the circumference is 176 mm ($\pi \times 56$ mm), to the nearest millimeter. Assuming that the tire's measurement is correct (and it wouldn't hurt to measure the diameter yourself), this tells us that the robot will move forward 176 mm for each complete wheel rotation.

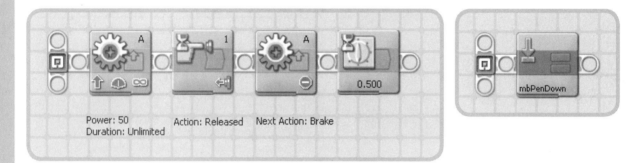

Power: 50
Duration: Unlimited   Action: Released   Next Action: Brake

Figure 16-5: mbPenDown and its icon

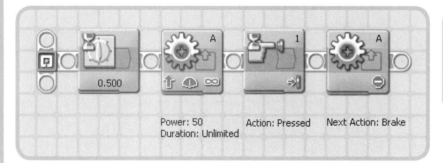

Power: 50
Duration: Unlimited   Action: Pressed   Next Action: Brake

Figure 16-6: mbPenUp My Block and its icon

To test this, we program Marty to move forward exactly 360 degrees (as shown in Figure 16-7) and then measure the distance actually covered. If our assumptions and calculations are correct, the distance we measure should be close to 176 mm.

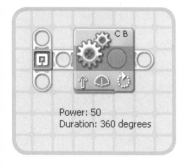

Figure 16-7: Moving forward 360 degrees

In Figure 16-8 you can see the results I obtained from running this command repeatedly.

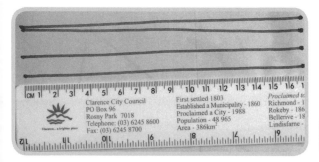

Figure 16-8: Four trials of moving forward 360 degrees

With measurements consistently around 170 mm, we can happily conclude that to move 170 mm (or 17 cm) we need to turn the wheels 360 degrees. From this we can determine the mathematical operations required to convert a given distance from centimeters into degrees (you will, of course, need to adjust this depending on your results):

$$170 \text{ mm} : \quad 360 \text{ degrees}$$
$$\Rightarrow \quad 17 \text{ cm} : \quad 360 \text{ degrees}$$
$$\Rightarrow \quad 1 \text{ cm} : \quad 360 \div 17 \text{ degrees}$$
$$\Rightarrow \quad X \text{ cm} : \quad X \times 360 \div 17 \text{ degrees}$$

This sequence means that we can multiply the desired distance by 360 and then divide that result by 17 to determine the necessary number of degrees. We can now program this sequence in NXT-G, as shown in Figure 16-9.

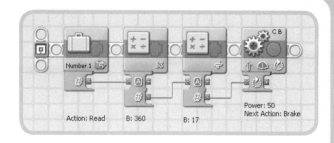

Figure 16-9: Preparing to create mbForward

NOTE To test this sequence you could omit the initial Variable block and instead type a desired distance into Input A of the first Math block.

Although watching Marty draw at speed can be quite breathtaking, you are likely to encounter some difficulties (as I did) when setting the Move block to a high power level. Specifically, I encountered a problem when drawing lines shorter than 10 cm. It seems that at moderate to high speeds, Marty will overshoot small distances and then back up to correct for them (as shown in Figure 16-10). Ordinarily, this probably wouldn't be a problem, but in our case it will result in a longer line than intended. For example, at a power level of 75, Marty consistently overshoots a 5 cm line by 1 cm. At a power level of 50 the overshoot is down to 3 mm or 4 mm. Distances of 10 cm and above seem to look fine.

Figure 16-10: Marty went too far and backed up.

The Variable block at the beginning of the sequence in Figure 16-9 is there to ensure that a data plug is created when we construct the mbForward My Block. To create the data plug, first select all but the Variable block(s), as demonstrated in Figure 16-11, and then create the My Block as usual.

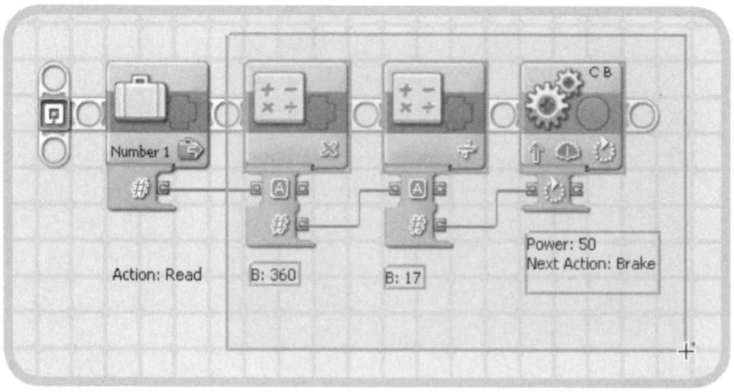

Figure 16-11: Selecting all the blocks except for the Variable block

Although you cannot add additional data plugs after a My Block has been created, you can change the description from the default Value, Value 2, and so on, as shown in Figure 16-12.

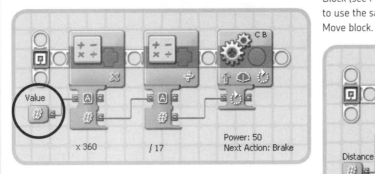

Figure 16-12: mbForward showing the generic Value property

To change the description of the data plug, edit the My Block and then use the Comment tool to edit the text above the data plug (see Figure 16-13). The modified description will now be visible in the block's data hub (see Figure 16-14).

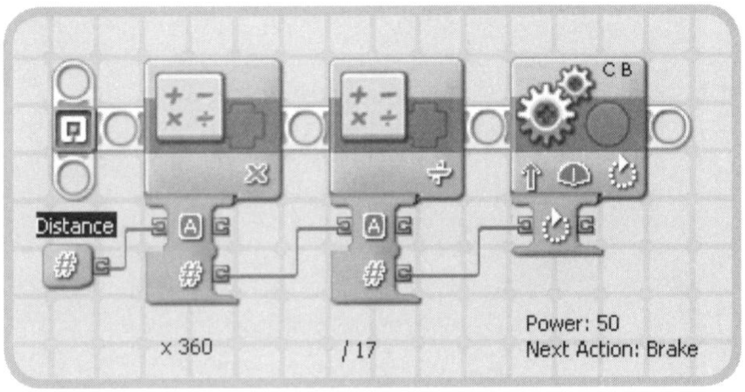

Figure 16-13: mbForward showing the renamed Distance property

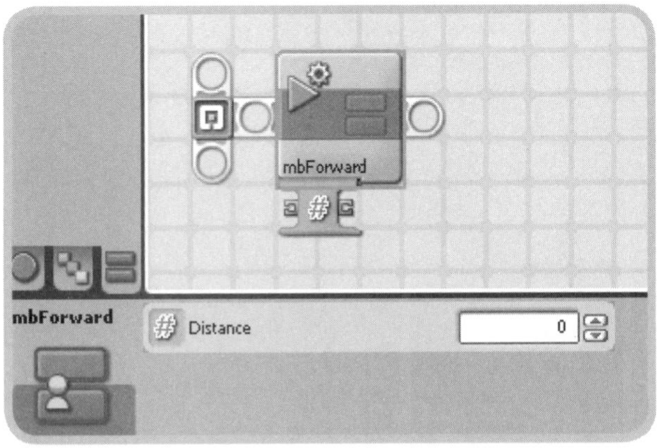

Figure 16-14: Icon for mbForward showing the Distance property

mbBackward

mbBackward

Having mastered the mbForward My Block, the mbBackward My Block (see Figure 16-15) is a piece of cake. The simplest approach is to use the same idea as mbForward, but change the direction of the Move block.

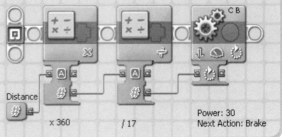

Figure 16-15: The mbBackward My Block and its icon

NOTE For something a little more sophisticated, you could scrap mbBackward entirely and extend mbForward to handle both directions by treating a negative distance as reverse.

mbTurnLeft

Not surprisingly, the process for creating the mbTurnLeft My Block (see Figure 16-16) is fairly similar to mbForward. Instead of moving forward a certain distance, however, we want Marty to turn through a certain number of degrees by having the two wheels rotate in opposite directions for a specific duration. The first challenge for this block, then, is to work out a method of converting any given turn angle into the number of the degrees that the wheels will need to turn through.

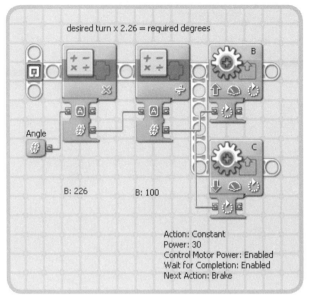

desired turn x 2.26 = required degrees

Angle

B: 226   B: 100

Action: Constant
Power: 30
Control Motor Power: Enabled
Wait for Completion: Enabled
Next Action: Brake

Figure 16-16: The mbTurnLeft My Block

Figure 16-17 shows how we can use a Move block set to a known duration together with some of the My Blocks that we have already created to draw a test angle.

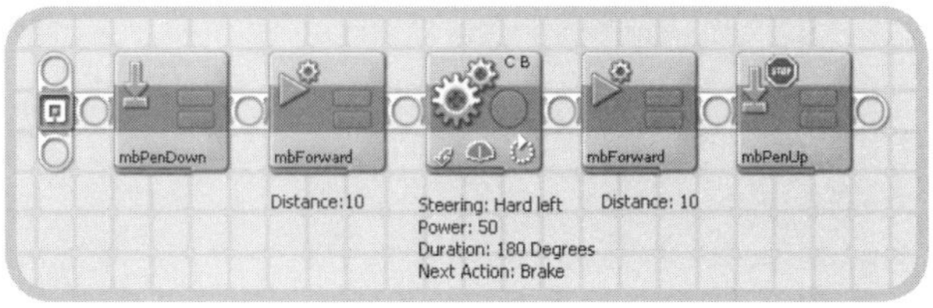

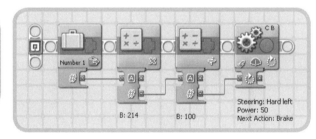

mbPenDown   mbForward   mbForward   mbPenUp

Distance:10   Steering: Hard left   Distance: 10
Power: 50
Duration: 180 Degrees
Next Action: Brake

Figure 16-17: Draw an angle using a duration of 180 degrees.

As you can see in Figure 16-18, the 180 degrees set in the Move block did not result in Marty turning through 180 degrees. In fact, the protractor shows that Marty turned through an angle of around 84 degrees, slightly less than half the 180 degrees set in the Move block.

NOTE The actual angle that the robot turns through will depend on factors such as the diameter of the tires and the distance between the wheels. Your angle will probably differ from mine.

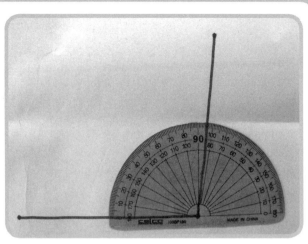

Figure 16-18: Results of drawing a test angle using 180 degrees duration

The results of our test tell us that to turn through a particular angle we need a Move duration of slightly more than double the desired turn angle. To be more precise: $180 \div 84 = 2.14$.

Unfortunately, the Math block in NXT-G does not allow us to multiply by a decimal fraction, so to work around this, we can first multiply by 214 and then divide by 100, resulting in the sequence of commands shown in Figure 16-19.

Number 1

B: 214   B: 100

Steering: Hard left
Power: 50
Next Action: Brake

Figure 16-19: Preparing to create mbTurnLeft

NOTE As with testing the mbForward sequence, you could omit the initial Variable block and instead type a desired turn angle into Input A of the first Math block.

Now there is just one more problem remaining to address, and again it relates to the way the Move block synchronizes the motors. The purpose of mbTurnLeft is to have Marty turn on the spot. For small angles it is not a problem, but for large angles, the result can be quite undesirable at times (see Figure 16-20).

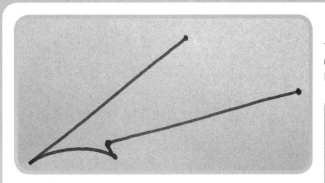

Figure 16-20: Not exactly Marty's tightest turn

mbTurnRight

The sixth and final My Block in Marty's basic toolkit, mbTurnRight (see Figure 16-22), is of course the same as mbTurnLeft with the Motor directions reversed. Too easy!

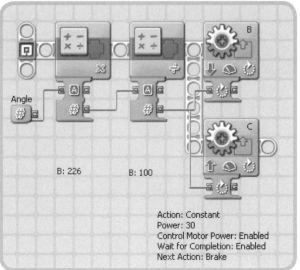

B: 226   B: 100

Action: Constant
Power: 30
Control Motor Power: Enabled
Wait for Completion: Enabled
Next Action: Brake

One way around this problem is to replace the Move block with a pair of Motor blocks (as shown in Figure 16-16). The motors won't be synchronized, but at least the errors won't be quite so obvious.

NOTE Both Motor blocks are executed in parallel to ensure that Marty does indeed turn on the spot. To create a parallel sequence beam, first drop the two Motor blocks onto the screen with one of them on the existing sequence beam; then hold the mouse over the sequence beam in front of the connected Motor block for a moment. You should then be able to drag a new beam to the start of the other Motor block.

Finally, in the same way that you modified the data plug of the mbForward My Block, change the description of the data plug to *Angle* (see Figure 16-21).

Figure 16-22: The mbTurnRight My Block and its icon

Now it is time to put these blocks to use to create some simple shapes.

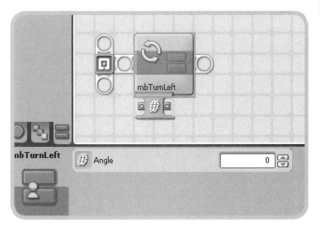

Figure 16-21: Icon for mbTurnLeft showing the Angle property

marty gets into shape

The following examples are intended to demonstrate some of the ways that Marty's basic My Blocks can be combined to make simple shapes. In some cases they are presented as My Blocks so that they can in turn contribute to more sophisticated drawings.

These examples only scratch the surface of what Marty can do. You might like to use some of these ideas as starting points or take a completely different approach. The only limit is your imagination.

mbPolygon

The next few programs, beginning with the mbPolygon My Block (shown in Figure 16-23), are designed to draw polygons. *Polygons* are defined as two-dimensional closed shapes, consisting of straight edges. Strictly speaking, these programs will allow Marty to draw only *regular polygons*; that is, polygons that have equal angles and equal side lengths (assuming that Marty behaves!).

Notice that the definition of polygon includes the word *closed*, which means that polygons have no gaps. Therefore, to succeed in drawing a polygon, Marty must finish where he begins.

NOTE If you consistently have problems drawing polygons, the most likely culprit is the conversion sequence within the mbTurnLeft My Block.

NOTE If you need to make changes to a My Block (such as mbTurnLeft), open it for editing by double-clicking the icon. Remember to save the My Block when you have finished working on it and then download the main program again. Also note that changes made to a My Block will be carried into any other programs that use that My Block.

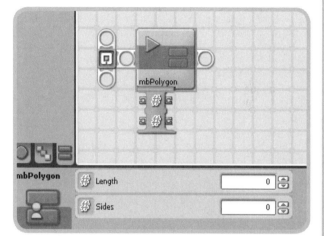

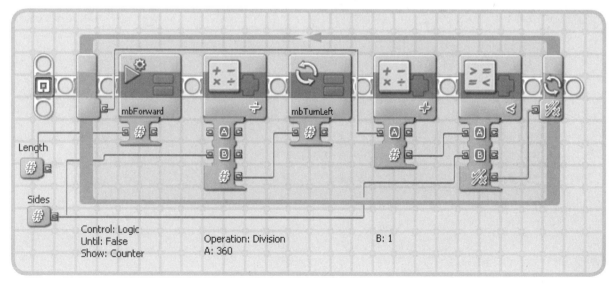

Figure 16-23: The mbPolygon My Block and its icon

If you spent some time experimenting with Marty's first program back in Figure 16-3, you might have already created a number of polygons (triangles, squares, pentagons, and so on). Although Figure 16-23 looks somewhat more complicated than Figure 16-3, the basic idea is the same: Move forward, turn, and repeat. Although this sequence would work fine as a normal program instead of as a My Block, the advantage here is that by being more general, it can be used in a range of ways, as the following programs will demonstrate.

length and sides

Notice in Figure 16-23 that this My Block has two inputs: Length and Sides. The Length input determines the length of each side of the polygon. The Sides input determines the number of sides of the polygon and is used in two ways. First, it provides a clever way of calculating how far Marty needs to turn at each corner based on the relationship between the number of sides and the turn angle required at each corner, as shown in Table 16-2.

table 16-2: the relationship between the number of sides and the required turn

name	number of sides	required turn (degrees)
triangle	3	120
square	4	90
pentagon	5	72
hexagon	6	60
heptagon	7	51.428 . . .
octagon	8	45
. . .	. . .	. . .

Notice that in each row of this table, the number of sides multiplied by the required turn is 360 degrees. From this observation we can determine the following:

required turn = 360 ÷ number of sides

We can also think of this in terms of how many degrees Marty must turn in total to end up facing in the same direction at the end as at the beginning: 360 degrees.

The Sides input is also used to determine how many times the loop should be repeated, so that the correct number of sides is drawn. Unfortunately this number cannot be wired directly into a Loop block, so the Loop is set to Logic with the count being compared to the number of sides each time at the end of each loop.

polygon

Now we begin to see how all the hard work of creating My Blocks pays off. Figure 16-24 shows how mbPolygon can be used to draw a polygon of any number of sides and any size, just by changing its settings.

NOTE We have an mbForward after the pen is raised because it's useful to have Marty move out of the way at the end of each drawing, so that we can see how he did.

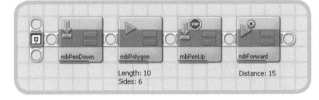

Figure 16-24: Polygon

Try changing the Sides and Length settings of mbPolygon and see how easy it is to create some different shapes, such as those shown in Figure 16-25.

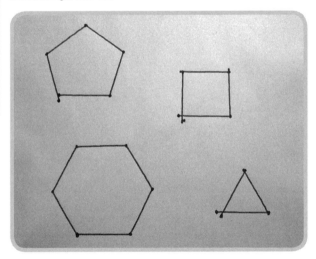

Figure 16-25: Examples of some different polygons

RandomPolygon

As a robot, Marty is naturally pretty good at following a set route to create the same drawings over and over again. But it's far more interesting to have Marty draw with a little more independence and spontaneity.

To introduce some unpredictability to Marty's repertoire, we can use some Random blocks. In Figure 16-26, Random blocks are used to determine the number of sides and the size of the polygon to be drawn.

Try running this program several times to see what will happen. Some of my results are shown in Figure 16-27.

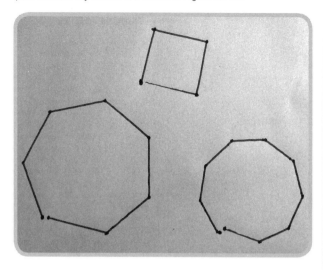

Figure 16-27: Examples of some random polygons

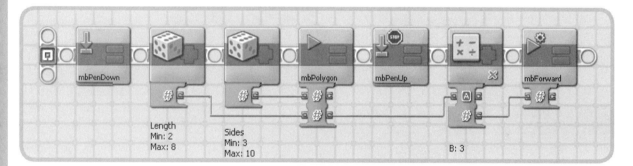

Figure 16-26: RandomPolygon

mbStar

No, Figure 16-28 isn't an accidental reprint of mbPolygon (refer to Figure 16-23), but a demonstration of how a small change to a program can make a big difference to the result. Can you see what the difference is here? It is a setting of 720 instead of 360 for Input A in the Math block.

From Marty's perspective, drawing a star is very similar to drawing a polygon: Move forward, turn, and repeat. From our perspective, however, the key difference is the amount of turn. By dividing the number of sides into 720 degrees instead of 360 degrees, Marty will effectively turn twice as far at each corner, resulting in his having completed a total of two turns by the time the shape is finished. The shape produced by this method is known as either a *star polygon* or a *polygram*.

NOTE This method of producing is only half true. Specifically, it works only when the Sides input is an odd number. I will leave it as a challenge for you to see how difficult it is to have Marty draw a star with an even number of sides. (Hint: The Star of David is an example of a star polygon with six sides.) If that's too easy, try to find a method for drawing even-sided star polygons and let me know (my email address is rob.torok@gmail.com).

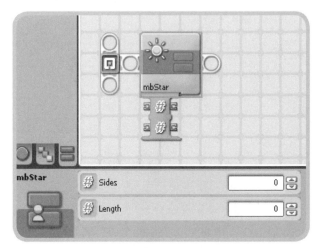

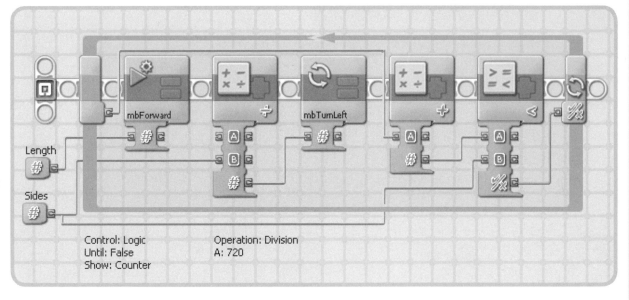

Control: Logic
Until: False
Show: Counter

Operation: Division
A: 720

Figure 16-28: The mbStar My Block and its icon

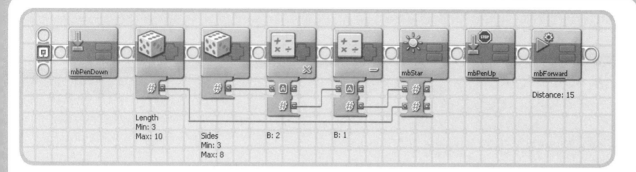

Figure 16-29: RandomStar

RandomStar

In much the same way that we gave Marty the capability to draw random polygons, we can also program Marty (see Figure 16-29) to create random stars (see Figure 16-30).

Figure 16-30: Examples of random stars

To force the number of sides to be odd, I have used the idea that one fewer than a double must be an odd number. Can you see why? Starting with a random number in the range of 1 to 8, we will generate odd numbers between 1 and 15, inclusive. To change the range of odd numbers produced, simply adjust the starting range.

mbZigzag

You might have noticed a slight bias to the left in all the examples so far in this chapter. To combat this, the program shown in Figure 16-31 introduces a more balanced shape. This program draws a zigzag by moving forward, turning one way, moving forward, turning the other way, and then repeating.

The three inputs are Steps, Angle, and Length. The Steps input determines how many zigzags are drawn. For example, a setting of 1 would produce a *V* shape, whereas a setting of 2 would produce a *W*. The Angle input determines how sharp the points of the zigzag are (the smaller the angle, the sharper the point). The Length input determines the length of each side of the zigzag.

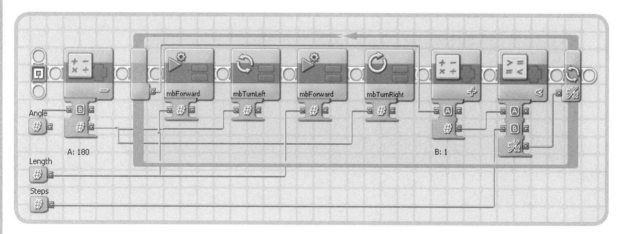

Figure 16-31: The mbZigzag My Block

Figure 16-32 shows the icon for mbZigzag, including the Steps, Angle, and Length properties.

Figure 16-34 shows some examples of running the RandomZigzag program (refer to Figure 16-33) several times.

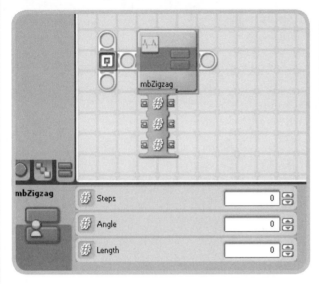

Figure 16-32: Icon for mbZigzag showing the Steps, Angle, and Length properties

Figure 16-34: Examples of random zigzags

RandomZigzag

Figure 16-33 shows how mbZigzag can be used to create a random zigzag. Experiment with the settings of each Random block to extend the range of zigzags produced.

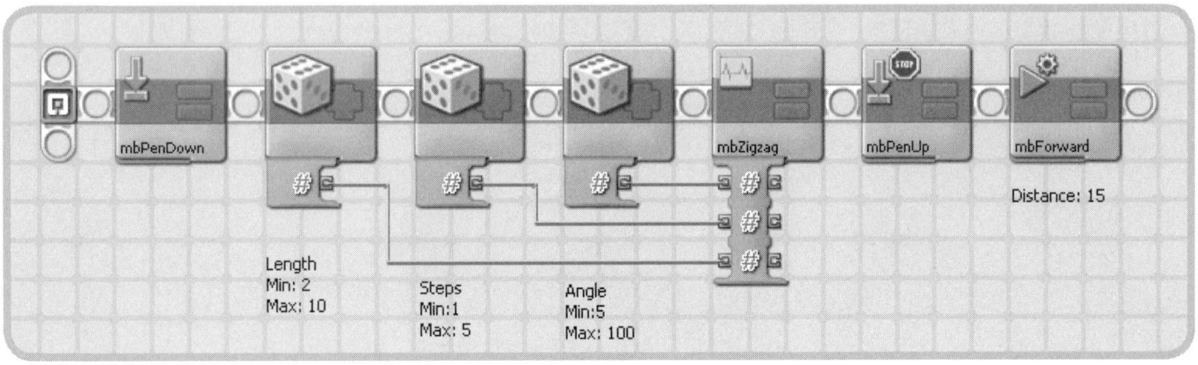

Figure 16-33: RandomZigzag

MegaRandom

Creating a random version of a particular shape is all well and good, but now we'll step it up a notch. The idea of MegaRandom is to combine all shapes that have been demonstrated so far in one program. As you add additional shapes to Marty's repertoire, you can further extend this program.

Figure 16-35 shows that the structure of MegaRandom is essentially just a Switch block inside an endless Loop.

By placing a different shape inside each option of the Switch block (as shown in Figures 16-36 through 16-38), Marty will continue drawing random shapes, separated by small gaps, until his batteries run out or his audience turns him off! If you have worked through each of the preceding example programs, you will not see anything new in Figures 16-36 through 16-38. In each case, I simply copied and pasted code directly from RandomPolygon (see Figure 16-26), RandomStar (see Figure 16-29), and RandomZigzag (see Figure 16-33), respectively.

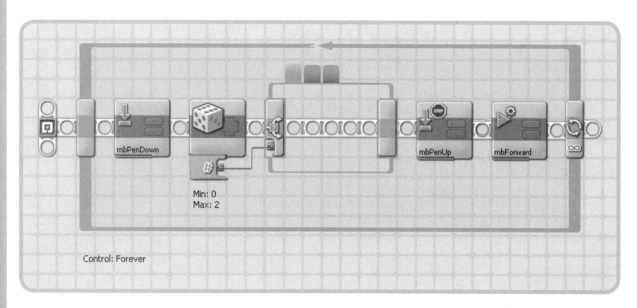

Figure 16-35: The basic structure of MegaRandom

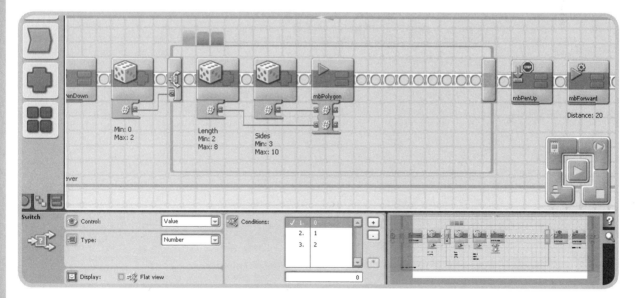

Figure 16-36: Inside the Switch block, option 1: Polygon

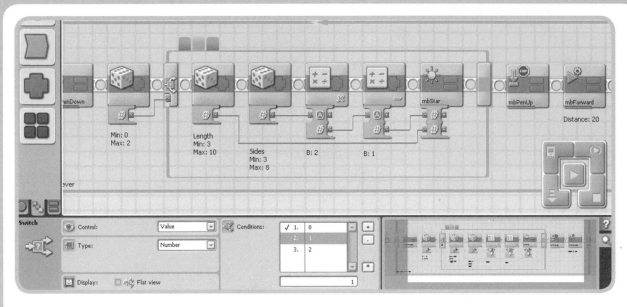

Figure 16-37: Inside the Switch block, option 2: Star

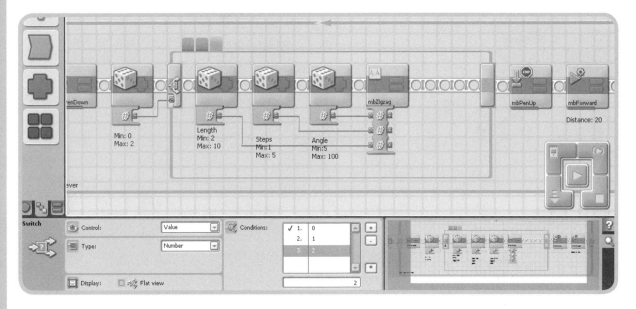

Figure 16-38: Inside the Switch block, option 3: Zigzag

Figure 16-39 shows the results when Marty runs MegaRandom.

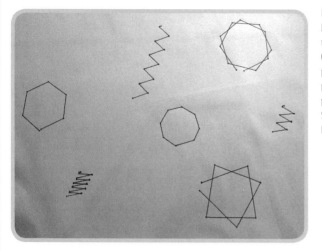

Figure 16-39: Marty, the performance artist!

SpiralStraight

Finally, here are two examples of how the counter of a Loop might be used to create a shape that develops according to the current value of the Loop counter. The first of these, SpiralStraight (Figure 16-40), is essentially a modified version of the RandomStar program, with a twist that each side is just a little longer than the previous one. The turn angle is chosen at random at the start of the program, but then remains the same for the rest of the drawing. Some examples from running this program repeatedly are shown in Figure 16-41.

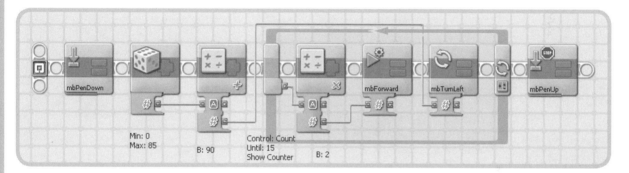

Figure 16-40: SpiralStraight

Figure 16-41: Examples of SpiralStraight

SpiralCurve

The final program I want to introduce, SpiralCurve (Figure 16-42), has no randomness in what it produces, but instead demonstrates that Marty is not limited to straight lines and on-the-spot turns. I leave it as a challenge to the reader to work out how using the value of the Loop counter to determine the speed of one of the motors can produce the result shown in Figure 16-43.

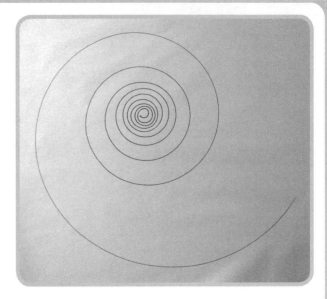

Figure 16-43: An example of SpiralCurve

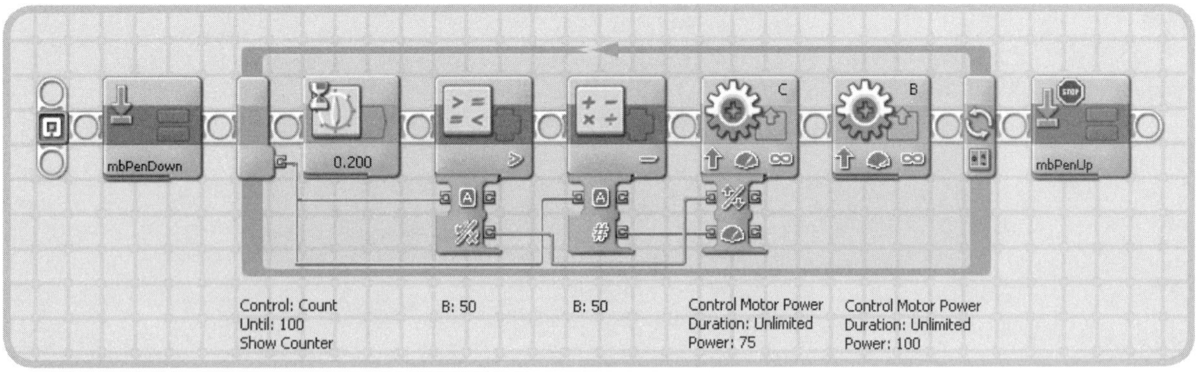

Figure 16-42: SpiralCurve

where to next?

I have offered only a small taste of what is possible with a performance art robot such as Marty. As I stated at the beginning of this chapter, my goal has been to show you how to write your own programs to create some original drawings. I hope that you feel confident about doing that now. I would love to see what you have Marty produce.

As you read through this chapter, I'm sure you've come up with some ideas for drawings that you'd like to add to Marty's repertoire. As for me, I'd like to give Marty the capability to do the following:

* Draw shapes based on curves, such as circles, waves, clouds, flowers, and so on.
* Use Sensors (e.g., Light or Ultrasonic) to stay within the drawing area. Like a floor gymnast, Marty could monitor his position to ensure that his drawing stays in a set area, such as on the paper!

* Use a text file to store the instructions for a particular shape. For example, a text file could include the lengths of lines to be drawn and angles to be turned to draw a picture or even letters of the alphabet.
* Further generalize the method for drawing a pointed star to cater to the cases in which a particular number of points can be joined as a star in more than one way.
* Use Bluetooth communication to have multiple robots (e.g., Marty and Martha) work together to create a drawing. Perhaps a group of robots could create a drawing while playing Follow the Leader.
* Use information from a range of sensors to influence what is being drawn. For example, the style of Marty's drawings could be influenced by audience applause, the amount of ambient light, or what he can "see" with his Ultrasonic Sensor.

troubleshooting

As with many artists, Marty tends to be a little temperamental at times. The difficulties shown in Table 16-3 are the most significant ones that I encountered as I was developing and testing Marty.

table 16-3: problems and solutions

problem	description and possible solution
Sloppy corners	The most common reason for a sloppy corner is that the marker pen is not exactly centered between the wheels. The marker pen should be centered in the pen grip, with respect to forward/backward as well as left/right. The shafts of the marker pens that I used with Marty were typically around 10 mm in diameter. You will need to adjust how the pen is held within the pen grip for different-sized pens. Sloppy corners will also result if either wheel is farther from the pen than the other, or if for some reason one tire is gripping the drawing surface better than the other.
Straight line wiggle and overshoot (refer to Figure 16-10)	Problems with drawing straight lines are usually the result of the way the motors are synchronized within a Move block. The problems tend to be more severe the faster the robot is traveling. Possible solutions include reducing the motor power or using a pair of Motor blocks instead of a Move block.
Inconsistent line thickness	Assuming that the pen isn't going dry, the most common reason for inconsistent lines is that the pen is too high. To solve this problem, slightly move the pen down within the pen grip. To some extent, the quality of the lines will depend on the pens you are using, but if the pen isn't low enough, you will find that the lines being drawn tend to be a bit sketchy. Perhaps it could be used as a deliberate technique for some drawings!
The pen keeps going up and down forever (yo-yo pen)	In contrast to the previous problem, the yo-yo pen occurs during the operation of mbPenDown if the pen is set too low and the pen grip mechanism actually lifts Marty slightly off the drawing surface, preventing the Touch Sensor from being released. The solution is to slightly move the pen up within the pen grip.

differences between sets

If you don't already have a MINDSTORMS NXT set, perhaps you're thinking you'd like to get one now. But what should you get, and where should you get it? There are actually two different versions of the MINDSTORMS NXT Base Set: the Retail version and the Education version. You can buy the Retail version at http://www.mindstorms.com, and the Education version can be found at http://www.legoeducation .com. This appendix describes the differences between the two versions and helps you figure out which one might be best for you.

why two?

You might be wondering why there are two different versions in the first place. This is because there are two LEGO divisions: the main Retail division and an offshoot called LEGO Education, which sells LEGO sets for use in schools. (Clearly, the NXT system is not only a lot of fun, but it is also educational.) The Retail version includes the accessories that LEGO thought would be best for most people, while the Education version is tailored for use in schools.

what do they cost?

Although both base sets are listed for $250, each set is missing components you'd need to buy in order to make just about any NXT robot.

One accessory you'll need to purchase if you choose the Education version is programming software. The Education version does not include any software, but you'll need to have it in order to make any substantial robots. You can buy the official software for the NXT system, called NXT-G, from LEGO Education for about $50.

If you buy the Retail version, you won't have to spend money on programming software because it's included in the set. However, the Retail version doesn't come with any batteries—unlike the Education version, which includes a special rechargeable lithium battery pack. You could buy six rechargeable AA batteries for about $25, or you could get the special lithium battery pack and charger from the LEGO store (http://shop.lego.com) for $73. Although expensive, the lithium battery pack is more durable than regular batteries and can be recharged without being removed from the NXT brick. This can be handy since it may be quite difficult to remove the batteries from some of the designs you come up with.

The least expensive combination that includes these accessories is the Retail version with AA batteries (about $275). Next would be the Education version and the NXT-G software, totaling about $300. The Retail version with the rechargeable lithium battery pack adds up to about $323. However, there are also other advantages and disadvantages of each version to take into consideration.

advantages of the retail version

Let's look at the basic advantages of each set over its counterpart, starting with the Retail version (Figure A-1).

Figure A-1: The MINDSTORMS NXT Retail version

I think of this version as having quantity instead of quality—the Retail version has a very large array of construction pieces. Along with the electrical parts, the Retail version comes with 519 building pieces chosen especially for constructing NXT robots. This gives it a big advantage over the Education version, which only includes 407 building pieces. Although you definitely need sensors and motors to make advanced robots, simple building pieces are essential as the backbone of your designs.

Another advantage to the Retail version is the test pad, which the Education version doesn't include. As you can see in Figure A-2, the *test pad* is basically a large sheet of paper with various marks on it (like a ruler, a black line, colored blocks, etc.) that are useful for testing your robots.

Figure A-3: The MINDSTORMS NXT Education version

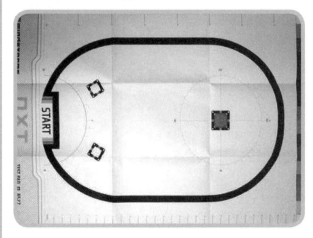

Figure A-2: The test pad

For example, suppose you made a robot with a Light Sensor and wanted it to detect blue areas of color. To do this, you'd need to know the reading given by the sensor when it's over a blue area. Since the test pad has a row of several different-colored blocks (including blue ones), you could simply put the Light Sensor directly over the blue square and view the sensor reading. The test pad also has a black line in an elliptical shape, which you can use to test something like a line-following robot. Altogether, the test pad makes for a great and easy way to test a wide variety of your designs.

advantages of the education version

Though lacking in quantity of construction pieces, the Education version (Figure A-3) makes up for it with quality, especially if you already have some spare building pieces lying around from other sets; it has a much bigger supply of electrical components.

Probably the most useful of these is the extra Touch Sensor. The Retail version only includes one, and to buy another one separately would cost $17. Touch Sensors are very useful for making bumpers, switches, and so on, and having two Touch Sensors instead of one lets you build robots that are more sensitive to obstacles or someone touching them. One example is a two-sided bumper (a bumper with Touch Sensors on both sides), which can not only tell your robot when it bumps into something, but on which side it hit the object.

Another nice addition to the Education version is the inclusion of three LEGO Lamps. (The Retail version doesn't include any lamps.) These are great for decoration or for making your Light Sensor more accurate. In order to use these lamps, however, you need a special type of cable called a *converter cable*, because the lamps were made to work with the older MINDSTORMS robotics kit, the RIS.

The NXT electrical parts use a different kind of cable than the RIS parts did. If it wasn't for these handy little converter cables, you wouldn't be able to use any of your old sensors, motors, or lamps with the NXT bricks. The Education version includes three of these cables, while the Retail version doesn't include any. As you can see in Figure A-4, one end of the cable has an NXT-type connector, and the other end has the RIS type. By attaching an RIS electrical piece to the RIS end and plugging the other end into the NXT brick, you can use an RIS piece with your NXT robots. This can be quite handy if you have many RIS sensors and motors hanging around that you don't want to go to waste. Converter cables not only let you continue using these old components but also save you the money you might otherwise spend on their NXT counterparts.

Figure A-4: A converter cable

which one?

Now you know about the differences between the two sets, but you may still be wondering which version you should get. The best version to buy depends largely on what you're willing to spend initially and which LEGO supplies you already have.

Suppose you already have some of the older RIS sets and pieces. In this case the Education version might be best for you. As discussed earlier, that version will let you continue to use your old RIS sensors (via converter cables) and give you a bigger array of electrical equipment. You can make up for the disadvantage of having fewer construction pieces by using the pieces from your RIS sets.

But what if you're new to LEGO MINDSTORMS and don't have an RIS set? If that's the case, the Retail version is probably your best bet. With its wide variety of construction pieces and a good amount of electrical equipment (though it's not as much as the Education version, it's still enough for most robots), the Retail version is a good starter set. Then, when you're ready to expand, I'd recommend getting the Education version with its greater supply of electrical pieces, since you'll already have a good amount of construction pieces.

There's one other situation I want to talk about. Perhaps you're like me: I have lots of old RIS stuff, and therefore have plenty of construction pieces. But when I build robots, I like to make them look nice, with uniform color schemes. I try to use two or three similar colors that blend well together, instead of having several colors jumbled together. The NXT pieces are mainly dark gray and white. Most of the RIS pieces, however, are black and yellow (along with light gray plates and some green pieces). These colors don't really mix well together, so if you're like me, you probably won't want to combine them if you can help it.

In this case, which set do you get? The extra construction pieces are still a factor, but it would sure be nice to have those converter cables and other exclusive features that come with the Education version. At this point, it really comes down to how much you want to spend.

If you're willing to spend a little more, I suggest getting the Education version along with an accessory set called the Education Resource Kit (Figure A-5), which you can buy from LEGO Education for $59. This kit includes 671 construction pieces made especially for use in NXT robots.

If you want to spend less, the best version to buy really depends on how badly you want to have uniform color schemes. If it doesn't matter too much, you may want to buy the Education version and put up with a few different-colored pieces, while getting all its advantages. If you really want to have uniform colors, though, then you'll probably want to go with the Retail version.

Figure A-5: The Education Resource Kit

B

trouble-free CAD installation guide

After you've spent many long hours creating a robot that you're really excited about, what do you do with it? If simply taking it apart and losing it forever is a painful thought, consider using the free software discussed in this appendix to digitize your work, whether to archive it or release it to the rest of the world. This is the same computer-aided design (CAD) software that we used to produce the illustrations in this book, and it's pretty amazing.

This is not an appendix on creating LEGO MINDSTORMS Robotics. Instead it is about what to do with your existing creations before moving toward designing something new and why you should consider using a computer to retain the wonderful inventions you spent long hours creating. If you have ever felt compelled to leave a model assembled as long as you can bear not to reuse the parts, this appendix is for you. It is about the importance of installing and using a specific collection of free computer programs created by some of the most truly generous LEGO fans in the community. While the developers are completely obsessed with producing extremely useful, high-quality software somewhat comparable to the awesome quality of the LEGO brand itself, there exist some shortcomings. The installation process can be a real nightmare (even using the All-In-One Installer available at http://ldraw.org). You will find an excellent installation guide in this appendix that facilitates the earliest stages of LEGO CAD knowledge and provides instructions for installing the core tools using the quickest and most trouble-free method ever written—thus, cutting the confusion completely.

why use the CAD programs?

Most of us have a limited number of real LEGO elements to utilize in our creative endeavors. This forces builders to dismantle one creation to build something new. Maybe the invention or construct is recorded with a camera before it is taken apart—or maybe not. Before these computer-based 3D rendering programs existed, this was my own process. It is nearly impossible to rebuild an exact reproduction, taking hours or days to closely re-create the same designs, and often never re-creating an exact copy of the perfectly working contraption again. This problem claimed several of my own designs—they're now lost and scattered among my parts bins. If your collection consists only of LEGO MINDSTORMS NXT set parts, you will find this appendix extremely useful. You will be able to build, render, and disassemble over and over again, without worry of a difficult reconstruction ever occurring.

just how good could this freeware be?

LDraw, the program at the root of these software applications, was created by James Jessiman in the late 1990s. Since his untimely passing, his incredible work has spawned the creation of several additional programs much coveted by today's users. The software is absolutely fantastic . . . unbelievably so, to be offered as freeware. However, working with these programs can be tedious and time consuming. As Fay Rhodes explains, "Creating the illustrations for this book has been a tremendous challenge. The programs are freeware and there is very little documentation for any of the programs we used. The only book written on the use of these programs is five years old—a very long time in the technology world."

what's the use?

If you are still wondering how using CAD will help you, look at the illustrations in this very book, as well as the abundant 3D renderings available on the Internet, and then sit back and discover that there really is no end to the possibilities.

The programs allow you to do the following things:

* Render whole models or even components for later use and easily, exactly, and quickly re-create that perfect design long after it was dismantled.
* Construct a limitless variety of LEGO models with unlimited amounts of virtual elements instead of real physical parts.
* Create, redesign, save/store, print, and trade step-by-step instructions (just like LEGO kit instruction manuals with building steps) on your personal computer.
* Teach other builders construction techniques, components, and mechanisms.
* Experience computer-based 3D graphic rendering concepts found in diverse fields such as architecture, engineering, and 3D animation and design.

RENDER RIGHT

Working with these programs can be tedious and time consuming, not because they're not sophisticated, but because CAD rendering tools are complex. Too, there is very little discernable documentation available for these programs, though you may find two (now out-of-date) books helpful:

* *Virtual LEGO: The Official LDraw.org Guide to LDraw Tools for Windows*, by Tim Courtney, Steve Bliss, and Ahui Herrera (No Starch Press, 2003)
* *LEGO Software Power Tools with LDraw, MLCad, and LPub*, by Kevin Clague and Miguel Agullo (Syngress, 2003)

While there are several different free LEGO CAD programs, the collection listed below will provide very good results. I'll help you to install each one later in this appendix.

* *LDraw*, where it all begins. (You will use MLCad to create the actual models.)
* *LDraw Complete Updates library*, the officially released LDraw LEGO elements.
* *MLCad*, the Windows-based LEGO CAD rendering and editing program that you'll use to create models, save snapshots and instruction pictures, and generate model parts lists.
* *LDView*, a small program that you'll use to view your models in fairly realistic 3D as well as save snapshots.

Please read through these instructions before you install anything, because it will be important to do certain things in a particular order. (We've left out specific program version numbers because version numbers will change over time.)

NOTE These applications are not official LEGO products, and this appendix will describe only how to install them under Microsoft Windows. Mac OS and Linux users will find information about freeware such as Bricksmith, LDGlite, and MacBrick CAD—along with easy setup instructions—in the "Get Started" areas at http://ldraw.org.

step 1: download and install LDraw

Go to http://ldraw.org/downloads.html and download the LDraw program (ldraw.exe) as well as the Complete Updates library (complete.exe) from the *Core Files and Libraries* link. Save both files to your root C:\ directory. (When installed, LDraw will create a subdirectory C:\ldraw.) Install LDraw by clicking the Windows **Start** button and selecting **Run**. In the box, type `C:\ldraw.exe -y` (or the appropriate filename for the version you've downloaded) and click **OK**. A pop-up window should appear as the files are extracted. When the program finishes, close the window.

a. Install the Complete Updates library by clicking the **Start** button and selecting **Run**. In the space, type `C:\complete.exe -y` and click **OK**. When the program finishes, close the window.

b. Now, you need to create the parts list (parts.lst). To do so, click **Start** and then select **Run**; type `C:\ldraw\mklist.exe` and click **OK**. When asked to sort by number or description, type **D** for *description*. When the program finishes, close the window.

step 2: download and install MLCad

Download MLCad from http://www.lm-software.com and extract the Zip archive to the new C:\ldraw\ directory. (The parts.lst file should be in this same directory.)

a. Next, tell MLCad where to find the LDraw parts library. Start MLCad by either opening the new icon on your desktop or choosing **Start ▸ Programs ▸ MLCad**.

b. A dialog box may appear, telling you the path is invalid, at which point the Settings dialog of MLCad will pop up, asking for the location of the root directory. Type `C:\ldraw\` and click **OK**. (You must specify the correct directory for MLCad to work! MLCad will not continue until the right path has been entered.) Once the right path is entered, MLCad will start up and show a start screen.

c. Select **File/Scan Parts** from the MLCad menu; MLCad should tell you that it has found new parts and ask if you want it to save a new parts.lst file. Click the **YES** button.

NOTE You'll find a very useful MLCad Tutorial at http://www.hpfsc.de/mlcd_tut/tut_eng.html.

step 3: download and install LdView

Download LDView from http://sourceforge.net/projects/ldview. Choose the newest release of the LDView.exe program file, and then save the file to the C:\ldraw\ directory. You can also create a shortcut icon on your desktop to easily start the program whenever you want to.

a. Run LDView by opening the LDView.exe file (or the icon if you created one). LDView will open its main window and immediately present you with an empty display (assuming you didn't specify a file as part of the launch process).

b. Select the LDraw/MLCad LDR, DAT, or MPD file you want to view; it should display the file in the window. (You can resize this window.)

NOTE If you open a large model within LDView, it could take a substantial amount of time to load. During this time, the status bar will indicate the progress of the load. Be patient, or else press ESC or choose Cancel Load from the File menu to cancel the load.

adding parts to the LDraw parts library

Can't find a LEGO part you need? Getting *Part Not Found* errors when opening files in MLCad? Most brand-new LEGO parts that have not yet been "officially" released for use in CAD programs can be found in the LDraw Parts Tracker at http://www.ldraw.org/library/tracker. Be sure to note which folder or subfolder each file should be saved to within your LDraw directory (see Table B-1 on the next page).

NOTE You should be able to find all the key NXT set parts at http://www.philohome.com/nxtldraw/nxtldraw.htm (which, as of this writing, had not yet been officially released by the LDraw committee). Follow the instructions there or below to install the parts.

official parts

If you need a replacement official part because you accidentally deleted or overwrote the file, just download the part from the

Official LDraw Parts Library using the URL here, where *<filename>* specifies the directory and name of the file within the LDraw library, according to Table B-1:

http://www.ldraw.org/library/official/*<filename>*

For example, to retrieve the current version of the 2 × 4 brick, you would type http://www.ldraw.org/library/official/parts/3001.dat.

NOTE Every time you add new parts library updates or a single unofficial part, subpart, or primitive, you will need to follow step b under "Step 1: Download and Install LDraw" and step c under "Step 2: Download and Install MLCad" for MLCad to recognize the new part(s).

unofficial parts

Unofficial parts might be incomplete or inaccurate, and it's likely that when they are officially released, they will change in ways that could mess up any model in which you use them.

The LDraw folks recommend that you grab unofficial parts one by one and use them only for specific projects. If you are using MLCad, you can import the unofficial part(s) as submodels into your main model files. That way, if changes are made to the unofficial parts before they are officially released, your model won't be adversely affected.

The *best* (that is, the safest) way to install files of unofficial parts is to unzip them to a temporary directory; then move them to your C:\ldraw\models directory. Don't dump them into your C:\LDraw\parts directory, because it will be very hard to find and remove them when the officially released part files become available.

When you unzip the files, be sure to tell your Zip tool to use the folder names, so that all the files end up in their proper subdirectories.

NOTE When installing unofficial files you might be asked if you want to replace files that already exist. Just say *No*. This means that the officially released files already exist, and you do not want to overwrite official part files.

parts library structure

The LDraw parts are arranged into certain folders so that the CAD programs can find, assemble, use, and render them efficiently. Table B-1 describes the various LDraw part file types and their specific subfolder placements within the C:\ldraw\ directory (which also correspond to the folder directories listed on each part description page found at http://ldraw.org). It is very important to place each file type into the correct subfolder when adding parts to your parts library. Using this table will enable trouble-free placement of parts into your parts library.

table B-1: LDraw parts directory general descriptions

directory	contents
C:\ldraw\parts	Complete, virtual parts, as used with rendering programs
C:\ldraw\parts\s	Subparts—reusable sections of parts used to render complete parts or for special use in model renderings
C:\ldraw\p	Primitives—little bits used to render finished parts and subparts
C:\ldraw\p\48	High-resolution primitives—high-quality primitives

conclusion

In this appendix you have learned the trouble-free method of installing an awesome set of CAD programs that will help retain all the wonderful inventions you have created. The reasons to use these programs are abundant, and learning to use them will be one of the greatest LEGO experiences you will have. Creating, sharing, teaching, and growing communities all have their basis in this subject. So if you are inspired to permanently add to your LEGO invention portfolio or cultivate the LEGO community by creating with these CAD programs and sharing healthfully, I encourage you to do so! The LEGO world is what we make of it.

index

about the website

The website supporting this book is located at http://thenxtstep.com. The site is centered on the popular blog The NXT STEP, which brings together news and information related to the LEGO MINDSTORMS NXT system. In addition to NXT-related news, you will find content supplied by the book's authors that you can download and use to further your explorations with the robots and ideas presented in this book. Through the website's forums, you can also submit your own pictures of and write-ups about the robots you've created based on the examples from the book (including modifications and upgrades). This site also provides access to the following areas:

Code Archive

We encourage you to follow the book's step-by-step NXT-G programming instructions. However, for added convenience, the NXT-G source files for each robot are available for download from the site. Just click the Books tab, follow the links for *The LEGO MINDSTORMS NXT Idea Book*, and follow the instructions to import code into your NXT software installation.

Book Updates and Errata

The Books section of the website has the latest information about known errors, whether they are typographical in nature or they relate to programming or building instructions.

The NXT STEP Forums

In the NXT STEP Forums (http://thenxtstep.com/forums.html), you can converse with the authors, find answers to technical questions, and access other discussions of interest related to LEGO MINDSTORMS NXT. If you would like to make comments or ask questions about *The LEGO MINDSTORMS NXT Idea Book* in particular, visit the book-specific forums page at http://thenxtstep.com/smf/index.php?board=9.0.

The LEGO MINDSTORMS NXT Idea Book is set in Chevin. The book was printed and bound at Malloy Incorporated in Ann Arbor, Michigan. The paper is Glatfelter Spring Forge 60# Smooth, which is certified by the Sustainable Forestry Initiative (SFI). The book uses a RepKover binding, which allows it to lay flat when open.

The Unofficial LEGO® MINDSTORMS® NXT® Inventor's Guide
by DAVID J. PERDUE

The Unofficial LEGO MINDSTORMS NXT Inventor's Guide teaches you how to successfully plan, construct, and program robots using the MINDSTORMS NXT set, the powerful robotics kit designed by LEGO. This book begins by introducing you to the NXT set and discussing each of its elements in detail. Once you are familiar with the beams, gears, sensors, and cables that make up the NXT set, the author offers practical advice that will help you plan, design, and build robust and entertaining robots. The book goes on to cover the NXT-G programming environment, as well as several unofficial programming languages that have been developed for the NXT set, providing code examples and programming insights along the way. Rounding out the book are step-by-step instructions for building, programming, and testing six complete robots, including "Bumper-Bot: The Exploring Vehicle" and "Claw-Bot: The Hunting Vehicle," that require only the parts in the NXT set. Includes an NXT Brickopedia and an NXT-G glossary.

OCTOBER 2007, 400 PP., $24.95 ($29.95 CDN)
ISBN 978-1-59327-154-1

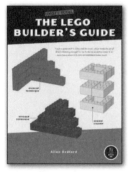

The Unofficial LEGO® Builder's Guide
by ALLAN BEDFORD

The Unofficial LEGO Builder's Guide combines techniques, principles, and reference information for building with LEGO bricks that go far beyond LEGO's official product instructions. Readers discover how to build everything from sturdy walls to a basic sphere, as well as projects including a mini space shuttle and a train station. The book also delves into advanced concepts such as scale and design. Includes essential terminology and the Brickopedia, a comprehensive guide to the different types of LEGO pieces.

SEPTEMBER 2005, 344 PP., $24.95 ($33.95 CDN)
ISBN 978-1-59327-054-4

Getting Started with LEGO® Trains
by JACOB H. MCKEE

Getting Started with LEGO Trains shows you how to build LEGO trains, from setting up train tracks to building custom freight cars. LEGO insider Jake McKee shares some of his most fascinating and original train designs, while including descriptive articles on basic building techniques and high-quality building instructions for several different projects. For veteran LEGO trains fans and curious beginners.

MARCH 2004, 120 PP., $14.95 ($21.95 CDN)
ISBN 978-1-59327-006-3

Forbidden LEGO®
Build the Models Your Parents Warned You Against!
by ULRIK PILEGAARD *and* MIKE DOOLEY

Written by a former master LEGO designer and project manager, this full-color book showcases projects that break the LEGO Group's rules for building with LEGO bricks—rules against building projects that fire projectiles, require cutting or gluing bricks, or use nonstandard parts. Many of these are back-room projects that LEGO's master designers build under the LEGO radar, just to have fun. Learn how to build a catapult that shoots M&Ms, a gun that fires LEGO beams, a continuous-fire ping-pong ball launcher, and more! Tips and tricks will give you ideas for inventing your own creative model designs.

AUGUST 2007, 192 PP. *full color*, $24.95 ($30.95 CDN)
ISBN 978-1-59327-137-4

PHONE:
800.420.7240 OR
415.863.9900
MONDAY THROUGH FRIDAY,
9 AM TO 5 PM (PST)

FAX:
415.863.9950
24 HOURS A DAY,
7 DAYS A WEEK

EMAIL:
SALES@NOSTARCH.COM

WEB:
WWW.NOSTARCH.COM

MAIL:
NO STARCH PRESS
555 DE HARO ST, SUITE 250
SAN FRANCISCO, CA 94107
USA